Impact of Information and Communication Technology on Libraries and Educational Communities

MJP
PUBLISHERS

Impact of Information and Communication Technology on Libraries and Educational Communities

A. Arogia Mary

Librarian in Auxilium College, Vellore.

P. Balasubramanian

Librarian of Manonmaniam Sundaranar University,
Tirunelveli.

MJP PUBLISHERS

Chennai **Tirchy** **Tirunelveli** **New Delhi**

MJP PUBLISHERS

ISBN 978-81-8094-362-1

MJP Publishers

Printed and bound in India

No. 44, Nallathambi Street,
Triplicane, Chennai 600 005

MJP 740

Publisher : **C. Janarthanan**

Project Editor : **C. Ambica**

LIBRARY AND INFORMATION SCIENCE

In partial fulfillment of the requirements for the degree of
DOCTOR OF PHILOSOPHY

MANONMANIAM SUNDARANAR UNIVERSITY

TIRUNELVELI -627 012

SEPTEMBER 2018

MANONMANIAM SUNDARANAR UNIVERSITY,
TIRUNELVELI - 627 012

CERTIFICATE

The research work embodied in the present Mini Project entitled "**Information Literacy through Madras University Library System, Tamil Nadu**" has been carried out in the Department of Library and Information Science, Manonmaniam Sundaranar University, Tirunelveli. The work reported here is original and does not form part of any other thesis or dissertation on the basis of which a degree or award was conferred on an earlier occasion to any other scholar.

I understand the university's policy on plagiarism and declare that the thesis and publications are my own work, except where specifically acknowledged and has not been copied from other sources or been previously submitted for award or assessment.

A. Arogia Mary
Research scholar

Dr. P Balasubramanian,
Deputy Librarian
Manonmaniam Sundaranar University,
Tirunelveli-627 012.

PREFACE

From the dawn of the last quarter of the 20th century, especially from the mid-1990s, Indian university libraries have made provision for electronic information resources and services. The main intention of making provision for electronic information resources and services was to improve the quality of library services for the benefit of academic community. The university libraries in Tamil Nadu are also making considerable efforts in this direction. The author wishes to conduct a study about the user pattern of electronic information resources in the University libraries in Tamil Nadu, because the present book focuses on the seamless usability of the electronic information resources and its growth, development and maximum use of these most modern electronic information are taken for a clear cut analysis and assessment. Today's research scholars may be more sophisticated than their predecessors in using some of the basic functions of many systems, however many are clearly unaware of the limitations and use of many electronic sources of information. Research scholars and faculty members entering research appear to have raised expectations about the capabilities of electronic resources. So this type of book is significant and important in the new IT Era.

ACKNOWLEDGEMENT

First and foremost I thank and praise **the Lord Almighty** who has been the source of wisdom and strength throughout my research endeavour.

It is my privilege to have the opportunity of working under the valuable guidance of, **Dr. P Balasubramanian,** Deputy Librarian, Manonmaniam Sundaranar University, Tirunelveli, Tamil Nadu, my guide who helped, enlightened and encouraged me in completion of this research study. I express my whole hearted acknowledgement with deep sense of gratitude and devotion and respect.

My sincere gratitude to our Honourable Secretary, Rev.Sr.Amalorpavam, S.J., M.A., M.Phil. and to our Dynamic Principal Rev.Dr.(Sr.) Reginamary R. of Auxilium College, Katpadi, Vellore.

I place it on record, my sincere thanks to my Uncle Rev.Fr. Gnanasekaran, S.J., Headmaster, Darsi, A.P. for his continuous help and encouragement.

I wish to thank the respondents, especially the Madras University Library users of study area of Chennai for their generous co-operation in providing me the necessary data and information in spite of their busy work.

I thank my husband Mr. A. James in a special way for his loving assistance and cooperation and my friends, for their support.

I am profoundly grateful to all my friends, relatives, colleagues and others who helped me directly or indirectly in the completion of this study.

I remember all those whose names are not mentioned here because no one is forgotten.

A. AROGIA MARY

LIST OF TABLES

LIST OF FIGURE

LIST OF ABBREVIATIONS

AACR	-	Anglo American Cataloguing Rules
ACRL	-	Association of College Research Libraries
ADONIS	-	Advanced Documents Over network Information System
ARIC	-	Agricultural Research Information Centre (New Delhi)
AUI	-	Attachment unit Interface
BIE	-	Book Index Entry
BLAISE	-	British Library Automation Information Service
CBC	-	Children's Book Council
CCC	-	Classified Catalogue Code
DMCA	-	Digital Millennium Copyright Act
ERIC	-	Education Resource Information Centre
FRBR	-	Functional Requirements of Bibliographic records
GILA	-	Government of Indian Library Association
HTML	-	Hyper Text Markup Language
IMCE	-	International Meeting of Cataloguing Experts
JATS	-	Journal Article Tag Suite
KWIC	-	Key Word In Context
LAMA	-	Library Administration and Management Association
MARBI	-	Machine Readable Bibliographic Input
NLM	-	National Library of Medicine
OAI	-	Open Archives Initiative
POPSI	-	Postulate-Based Permuted Subject Indexing
QDL	-	Quick Development Library

CONTENTS

CHAPTER - I

Introduction

1.1 PREAMBLE

The information produced in the world before 1990"s was available in print form and a part of information availability was in the non-print form such as audio, visual sources, microfilm, etc., and were considered non-book materials. The information world has witnessed a transition from print to electronic form. A change would not happen, all of a sudden but it happens gradually. When a medium of publication changes, it requires time. It took around ten years and it still needs a couple of years. Now people predict that in the next few years most published information will be available only in the electronic form. Hence, we need to understand and use the electronic media to the maximum extent.

The electronic publications in the last decade have increased dramatically to the extent that no one can propose equations to reflect this reality. Any exponential increase is characterized by the lack of structure, organization and control. Electronic publications are ideal examples of the above properties. The major individual electronic forms are internet, CD-ROMs, online information and magnetic tapes. Among various forms of electronic publications, probably the www is the most predominating one. The www is an unorganized and unstructured distributed electronic information.

Advances in computer network and storage technologies have inspired the design of digital libraries in recent years. The emergence of digital libraries in recent years has introduced a number of important issues. One issue that has not attracted much attention, but is essential to digital library development, is the management of digital library contents because a typical digital library contains a large quantity of documents, to the conventional practice used to maintain. The contents of websites will not work for digital libraries. In particular, the hyperlink that is widely incorporated in conventional websites to structure the web pages pose a major challenge. For a digital library with large quantity of materials attempting to manually maintain the hyperlinks among web pages will become a cumbersome task for the manager of a digital library.

Library exists in many forms and are of many types. With multimedia technology have image libraries, audio libraries, and even digital video libraries. We might also think of libraries when we refer to collections that now reside in database, text bases and the World Wide Web (www). Traditional library budgets devote an ever increasing share of their funds to electronic services, whether in the form of CD-ROMs, Online Public Catalog (OPAC) or Online Database.

As digital storage costs go down relative to the cost of library shelf space, and as electronic services become more useful, affordable, available and usable, there is a move towards the grand challenge of the World Digital Library System. In the digital library, users will need to work simultaneously with multiple distributed information sources that differ in content, form and sources types. They will need to understand displays of larger sets of objects from these sources, organized in ways that allow them to make rapid decisions about what is relevant and what overall patterns exists among the objects.

Libraries serve at least three roles in learning. They serve practical role in sharing expensive sources – both physical and human resources. Libraries serve a cultural role in preserving and organizing facts and ideas. Libraries serve as centers of interdisciplinary places shared by learners from all disciplines. Digital libraries extend such interdisciplinary approaches by making diverse information resources available beyond the physical space shared by a group of learners.

1.1.1 Library Network

An information network is a set of interrelated information systems associated with communication facilities, which cooperate through more or less formal agreements and institutional agreements, in order to jointly implement information handling operation, with a view to pool in their resources and to offer services to users. They generally follow identical or compatible rules and procedures.

Library Network Components

A library network consists of six major components

- Information resources
- Readers or users of information
- Schemes for intellectual organisation of data/documents
- Methods for the delivery of information to the user- ie the output
- Formal organisation – cooperation, whether voluntary or obligatory, in information exchange and utilisation.

- Bidirectional communication links based on these components, the
- essential requirements of INFLIBNET can be identified as Libraries (University, R&D Institutions, etc.)

1.2 ELECTRONIC INFORMATION ENVIRONMENT

An electronic environment allows changes and updating of original information, provides different views/reading of the same document, integrates multimedia sources of information, permits interchange of data and offers software support online. All these facilities while being useful for some types of publication are not appropriate for every kind of book; different kinds of reading requirements make electronic translation more or less useful for the reader.

Features

Provide access to very large information collections including access to primary and complete information, not merely surrogates or indexes.

- Support multimedia content
- Network accessibility
- User friendly interfaces
- Unique referencing of digital objects
- Multi use refer various area in same time
- Current information science
- Advanced search and retrieval
- Supporting both formal and informal learning
- Remote access
- Online discussion& commands
- Accessibile to anyone, anywhere, anytime, during travel, hotel etc. is supportive, opportunity to publish, annotation and integration of new information.

1.2.1 Electronic Information Resources

Any information resource that is accessible through computers or network can be termed as electronic resources. It is also available through the Internet or through online databases. Electronic Information Resources (EIRs) originally published information in electronic form or in print form made available electronically.

Electronic media has proved its advantages over the print media. Any information resource that is accessible through computer or network can be termed as electronic resources like e-journals, e-books and e-data bases that have increased considerably.

Computer storage devices such as optical disk, CD ROM/DVD-ROM Databases accessible through Internet and other networks can be used or stored for further use.

Electronic Information Resources (EIR) have their origin in experimental computer Systems developed for the storage and retrieval of bibliographic data in 1960s. By the end of that decade some of the major bibliographic databases such as Chemical Abstracts and Index Medicus were available in magnetic tape versions that were searchable in offline batch mode. During the 1970s and 1980s, the increasing availability of this machine-readable data together with the emergence of both real-time interactive computing and computer networks enabled the online information industry to emerge. Initially the major scientific bibliography databases became available in machine- readable form. During the 1980s, Academic libraries began to transfer from card catalogues to online public-access catalogues (OPAC) OPACs became widely available; the CD-ROM emerged as an information delivery vehicle. The emergence of the World Wide Web has enabled a revolution in electronic information resources. This difference from the earlier situation is that:

- The available information is not restricted to text but includes large numbers of images, audio and multimedia items so that it is more appropriate to think of them as information objects rather than documents.
- The information available is an amorphous mass to which anyone can add if they have even a limited knowledge of Hyper Text Markup Language and thus the available information is no longer subject to quality-control mechanisms prior to publication.
- The information is not structured to facilitate retrieval, but through the hypertext links it is structured to facilitate browsing and easy moving between information objects.

The web browser environment has continued the trend towards user friendly interfaces that was initiated by the development of CD-Rom and windows.

Alongside considerable change in the type and scale of available EIR, there has been an even more remarkable change in the users of these resources. Use of EIR has moved from being an esoteric activity undertaken by information professionals, and a slowly increasing band of other professional people, to an action-undertaken everyday by countless millions around the world.

Electronic Information Resources are a long and involved process and should be carefully planned. Librarians who carefully weigh the experiences of the libraries offering electronic resources and who consider and prioritise selection criteria, will make the widest choice and will have a program that makes the transition to electronic resources with relative ease. As the library begins the selection process, one of the first steps to be taken is to develop a list of the kinds of products the library will be first purchasing. After that, there are literally thousands of products that can be purchased. Electronic databases greatly increase access to information and the implementation must be carefully planned to ensure optimal use of space, money and time.

1.3 E- RESOURCES

E-Resources are those electronic products that deliver a collection of data, be it text referring to full text basis, e-journals, image collection, other multimedia products and numerical , graphical or time based , as commercially available till that has been published with an aim to being marketed. These may be delivered on CD ROM , on tap via internet and so on. The library and information services of the 21st century are fast changing. With the rapid development on electronic publishing , libraries are not only acquiring reading materials such as printed books and journals but also provide access to various learning resources in the electronic form. In the new situation , the role of library professionals in collection development and management has become restricted.

1.3.1 Components of E- Resources

1.3.1.1 CD-ROM Databases

CD-ROM databases publications are increasing day by day in almost all fields due to their variety of advantages in information storage and retrieval. Majority of the publishers of books and journals, on-line vendors and various learned societies are bringing out new titles in the CD format with a powerful search capability and user- friendly retrieval software. The CD-ROM technology has given ample opportunities for information professionals to introduce more information services to end-users. Electronic information resources in the CD format includes abstracting and indexing services Encyclopaedias, Dictionaries, Directories, Year-books, Back Volumes, Patents, Standards and many other reference works.

1.3.1.2 DVD-ROM Databases

The advent of the DVD-Digital Video Disc or Digital Versatile Disc, with its 17 GB of high data storage capacity, has made it possible to include more multimedia elements like video, audio and animation to integrate many reference sources on a single disc. The other features like higher quality of sound and video, higher rate of data transfer and data security, etc., are making DVDs a more viable option than CD-ROM. But, at present due to problems like lack of standards among the manufacturers of DVDs and drives, need for extra hardware on PCs and their higher prices are making the growth of DVD technology slow[4]. Some DVD reference sources include Britannica DVD 99, Webster"s International DVD Encyclopedia-2000, Grolier Multimedia Encyclopaedia, and Eyewitness World Atlas DVD-ROM Deluxe Edition, the complete National Geographic on DVD-ROM, etc.,

1.3.1.3 Online Databases

The recent growth of Internet and the popularity of ease, in use of WWW are making libraries subscribe online information services. The online database services like Dialog (KR information) and STN now move towards being webcentric. The usage of online databases against their CD-ROM counterparts has to be evaluated and decided on the basis of cost effectiveness and timeliness. A few Online information services are KR ScienceBase and STN. The KRscienceBase, includes information sources like BIOSIS, CA Search, Elsevier Science Publishers, Reuters, NTIS, etc., The STN international provides a complete collection of in-depth databases in science and technology which give quick, direct links to literature, patents and chemical catalogues. Chemical Abstract Services, the producers of the World"s largest and most comprehensive database of chemical information, offer several databases on STN like Caplus, INSPEC, COMPENDEX, MEDLINE, etc,

1.3.1.4 E-Journals

E-Journals or Electronic Journals are gaining more importance with the emergence of Internet. The Publishing world is undergoing a revolutionary change as more and more publications are becoming WEB centric. It is also accepted and stated that "most of the publishers are choosing WEB as an access medium using HTML to markup the journal content so that it can be read using a web browser. These are also called networked E-Journals. Other publishers have chosen to use additional access software in conjunction with the WEB, the most popular being Adobe Acrobat and its associated Portable Document Format (PDF) file type".

1.3.1.5 E-books

Electronic books include the hardware, a suitable device to read electronic media, perhaps better called „e- book reader". The hardware is important as it provides what readers may need to exploit with the software available and link this to specific requirements. E- Books help readers with more possibility of access and media of learning. They can be accessed from anywhere at any time and are free from time lag. The books need to be compatible to all hardware devices. They need to support any format to make it easy to transfer and read. The features of e-books include; their portability that gives a new chance of learning, e.g., distance learning. It can be carried and transferred anywhere, their feature likes changeable font size makes it easy to use and searchable and navigable through links provided.

1.3.1.6 E- Newspapers

Electronic news resources like lexisNexis and Factiva links to local, national and international newspapers.

1.3.1.7 E- databases

E-databases include periodical indexes and abstracts. Library and Information Science Abstract (LISA), WIPS (Worldwide Intellectual Property Search), METADOX (Metal Abstracts), SCOPUS (Abstracting , Indexing and citation data in science , Engineering and technology. Emeralds, EBSCO, are some of the examples of online database.

1.3.1.8 E-thesis

Electronic theses creation, use dissemination, and preservation of dissertations. We support electronic publishing and open access to scholarship in order to enhance the sharing of knowledge worldwide. e.g Network Digital Library of thesis and dissertation, Vidyanidhi Digital Library & e-scholarship portal.

1.3.2 Benefit of E-Resources

1. *Libraries*

- More content/ service
- More Value for Money
- Save Shelf Space
- Higher Internal Efficiency

2. *Researchers*

- Easier to search
- Easier to cite
- Convenience to access

3. *Authors*

- Wider distribution
- Increased usage
- Global readership

4. *Other benefit of e- resources*

- Quick access
- Remote access
- Mobility
- Space consumption
- Research comfort ability
- Multiple accesses
- Job simplicity
- Acquisition and maintenance

1.3.3 Types of E-Resources

Broadly three types of E-Resources are available to cater scholarly information needs of users:

1.3.3.1 Bibliographical E-Resources

These are the resources that provide bibliographical information like: title, author, name of the journal, volume and issue number, publication date, publisher and most importantly an abstract of the entry. Bibliographic e- resources mainly contain information on single a subject.

1.3.3.2 Full –text E-Resources

These E-Resources provide full- text of the document apart from its bibliographical information.

1.3.3.3 *Indexes*

An index is a reference source which provides bibliographic information about journal articles, as well as other types of materials. While indexes have long existed in print, online indexes have expanded the type of work done by researches. There are more options than looking for materials by subject, author or title. Online indexes allow you to look beyond subject, author, or title. They allow you to look for keywords or phrases throughout the bibliographic information including the abstract.

1.3.3.4 *Library Catalogs*

Most libraries now provide access to their catalogs from their web sites. Many others provide information about their holdings into larger databases such as WorldCat or the RLG Union Catalog. The Library provides links to these catalogs under the "Catalogs" section on its web site.

1.3.3.5 *Reference Sources*

Many dictionaries, almanacs, encyclopedias, and other reference sources are now available online in full-text.

1.3.3.6 *Statistical Sources*

The Library has access to a variety of subscription databases which provide economic data or statistics.

1.3.3.7 *Sound Recordings*

There are only a few Library databases which provide access to sound recordings. If you are looking for music online, start at the Music Subject Guide for the resources.

1.3.3.8 *Image Databases*

(Art, Maps, Medical, etc.): Some databases include graphics or images, such as photos, paintings or maps.

1.3.3.9 *Portable products*

Portals provide a single interface to search various database, e.g., Under UGC INFO NET: E-journals consortium Indian Universities are accessing more than

20+ different database or Products. Users have to search individually for a single term if they want to search in all the provided products. This exercise needs much time to cover all 20+ products.

1.3.4 Use of E-Resources in Libraries

Electronic resources are becoming more and more available in libraries. The print media is now being digitized, which increases the availability of books and journals in the electronic format. Electronic books are helpful because of their portability and feature of incorporating more than one book in a single hand held device. The published material is also available on open access. This also helps poorer people to access information free of cost. They need not worry about licensing and usage of the information.

13.5 E-Resources in the Libraries in India

The potentials of ICT has spread over in several ways to bring development in all academic institutions. Some of the programmes in India are listed briefly.

- Media Lab Asia aims to carry out the development of ICT tools and solutions, and generate in properties for gainful exploitation towards deployment of effective ICT based products for solutions.
- Development of Gateway Foundation (World Bank) Research and Development training centre concentrates on the use of internet, language and speech technologies, for the socially relevant sectors of Healthcare, Agriculture and Education.
- Vidhyavahini connects a pilot scale of 140 schools in 10 districts spread across the country providing learning material as the district education mode on interactive basis.
- Digital Library enables e-collections of literature with facilities for indexing and search from a large repository.
- The UDL project (Million Book Project) aims at creating a free-to-read, searchable collection of one million books available on the Internet.
- Mobile Digital library allows school children in rural areas to access Internet via a VSAT connectivity to download and take print out of popular books or other information of their interest.
- BSNL"s Internet Dhabas provides Internet connectivity from PCOs covering almost 5,00,000 Villages.

1.4 ELECTRONIC INFORMATION RESOURCES IN THE LIBRARIES IN TAMIL NADU

Tamilnadu is one of the constituent states of Indian Republic. It is the fifth largest state in India and it is the first state to enact legislation for library provision, doing so as early as 1948. There are a number of academic, public and special libraries. The late 1990s saw the introduction of IT in libraries, though mostly in academic and special libraries. The emerging national networks, namely, Information and Library Network (INFLIBNET), Development of Library Network (DELNET) and other regional and local networks such as Madras Library Network (MALIBNET) have all helped academic libraries in India in general and Tamilnadu, in particular, adopt IT facilities. It could also be surmised that the financial support of the University Grants Commission (UGC) of India is also likely to have been a factor. Special libraries in Tamilnadu are dynamically involved in the modernization of their libraries, though these libraries have built up their collections mostly by means of printed resources in combination with smaller quantities of digital materials.

The libraries in Tamilnadu can be categorized by type and by number as follows: Universities Libraries (25), Arts and Science colleges Libraries (290), Engineering Colleges Libraries (233) and Public Libraries (2,846). In addition to the above there are a large number of special libraries attached to various industries, research institutes and other corporate sectors. These libraries are slowly marching towards the automation of their services.

1.4.1 University Library

"The role of libraries in providing widespread and inclusive access to knowledge is widely acknowledged. In today"s context, libraries have to play two distinct roles – to serve as acknowledge centre of information and be a local gateway to national and global knowledge centre of information. In order to achieve this goal, existing libraries must modernize their collection, services and facilities, and become more proactive and collaborate with other institutions, agencies and NGOs in order to develop a community- based information system.

The Radhakrishnan Commission (1948) on University Education in India aptly observed that "The library is the heart of the university□s work, directly so as regards its research work, and indirectly as regards its educational work. Scientific research needs a library as well as its laboratories while for humanistic research, the library is both the library and laboratory in one." Library has thus become the heart and soul of university system. Collection of information resources is an important activity of any library. Collection development is a complex process that includes assessment of user information needs, evaluation of present collection, designing collection development policy, procurement of information sources and planning for resource sharing. The university library takes into account the information needs of users - teachers, students, researchers, etc. and develops the collection of all types of information sources in whichever format they are published.

Till recently, the libraries were procuring information sources in print form and organizing them physically in the libraries. The developments in ICTs have changed the world of information industry. The adoption of electronic technology in production of information sources has made collection development in libraries a more complex process. The developments in Internet services in general and WWW in particular have changed the face of library and information services in higher education and research.

The libraries are providing access to the information through computer networks, which have broken down the barriers of space and time. Hence, the shift is from information ownership model of librarianship to the information access model. The information behaviour of library users and their expectations of library and information services have also changed.

1.4.2 E- Resources and Services in University library

Electronic information resources offer unique advantages: compact storage, rapid retrieval and delivery, interactivity and flexibility in the transfer of information. These sources provide access to current information as these are updated frequently. Often, the sources such as full-text databases, e-journals, e-books including reference sources (dictionaries, encyclopedias, directories, handbooks, atlas etc.,), text-archives, OPACs, image collections, multimedia products, collections of numerical data, etc., provide links to related information to the users in their field of interest. The university libraries have realized the importance of electronic information resources to fulfill the ever-growing and dynamic needs of their new generation of users, who give more importance to instant access to information than its authenticity and authoritativeness. They are Internet savvy.

Internet is considered as the treasure house of information which provides access to information resources both free and fee-based. Internet has become ubiquitous all over the world, and university libraries are no exception to it. University Grants Commission established INFLIBNET Centre in the year 1991, to help the university libraries in India not only to automate their house-keeping operations but also to establish campus-wide networks in the universities.

The consortia activities, such as UGC-Infonet, INDEST and others are gaining popularity among higher education and professional institutes in India. By making use of all these electronic infrastructure and resources, university libraries provide different kinds of information services to their users such as: information alert services, news groups, virtual reference service, retrospective search service, document delivery service, Web-based instructions (FAQs), translation service, user education, institutional repository and portal services.

University libraries can also alert their users of information resources procured recently, forthcoming conferences and seminars, important information appeared in newspapers, newsletters, etc. Personalised services are provided to the individual users by passing information of their interest to their desktops. When the libraries introduce new resources and services all the users are informed of the list services. The university libraries create news groups and post them for wider dissemination among the users online to obtain their participation, suggestions, feedback, etc., regarding various resources and services offered by them.

University libraries offer virtual reference services to their users through email, instant messages (chatting) and over telecommunication. General guidance in making use of resources and services are provided through FAQs. University libraries have made provisions for automated circulation, reservation and recommendation for procurement of information sources, and access to the OPAC which saves the time of the academic staff. A well-designed library web site serves as a gateway for the resources of the library and promotes the use of information sources and services. The information sources which are more important are mounted on the library website in order to bring to the notice of the users, and their instant access.

Through robust and reliable campus network, university libraries provide attractive services to the academic staff. Information about the subscription of online sources and information sources available through consortia is passed on to the academic staff"s desk-top. The CD-ROM server connected to the campus LAN provides access to the information sources available on CD-ROMs and DVD-ROMs at the door-steps of the academic staff.

User education is another important service provided by a modern library. University libraries arrange training programmes, workshops, special lectures, and audiovisual presentations; publish brouchers, handbooks, and tutorials on computer and Internet fundamentals, searching CD-ROM, DVD-ROM, Internet, consortia-based resources, institutional repositories, etc.

1.5 STATEMENT OF THE PROBLEM

Every institution engaged in pursuit of higher education and research is trying to achieve academic excellence. Every educational institute has realized the need for automation of their libraries. The university libraries in Tamil Nadu are spending lakhs of rupees every year for the purpose of library automation and procurement of electronic resources. The UGC is also providing financial assistance for this purpose. Further, the UGC, under its UGC-INFONET programme, is providing access to e-journals to almost all the Indian university libraries for the benefit of the academic community.

There seems to be no definite answer for several research questions like – the strategy being followed by the university libraries in Tamil Nadu with regard to searching pattern in electronic environment by research scholar, collection of electronic information resources, the awareness of electronic information resources and services available in university libraries by the Research scholars and such other questions. A review of the published literature in the selected area of study reveals that, no systematic study has been carried out either on the availability of ICT infrastructure, and e-resources and services, or on their awareness, searching pattern and use by the users of **Madras University** library, of Tamil Nadu. And hence, the problem of the study is stated as: "Information Search Pattern in Electronic Environment by Research Scholars at Selected **Madras University in** Tamil Nadu: A Study."

1.6 NEED AND SIGNIFICANCE OF THE STUDY

From the dawn of the last quarter of the 20th century, especially from the mid 1990s, Indian university libraries have made provision for electronic information resources and services. The libraries are equipped with a number of computers, Internet connectivity, Email facility, Local Area Network (LAN), subscription of online resources such as e - journals, e-books and e-databases, access to consortia-based resources (such as UGC - INFONET), and CD-ROM databases. The automation of their traditional services like acquisition, cataloguing, circulation, serials control, etc., have been carried out. The main intention of making provision for electronic information resources and services was to improve the quality of library services for the benefit of academic community. The university libraries in Tamil Nadu are also making considerable efforts in this direction.

The large amount of public money is being invested for this purpose. Though the university libraries are non-profit organizations, there is a need to study the extent to which the electronic resource-based services have added value to the higher education and research.

There is an urgent need to know the information searching pattern of research scholars in electronic environment, their awareness about e-resources, strength and weakness of the collection of electronic information resources and services available in Tamil Nadu State university libraries. It is also because that, this study would reveal the problems faced by the research scholars while accessing electronic resources.. This study becomes an eye opener with regard to the existing situation of information searching pattern of research scholars in university libraries of Tamil Nadu. It is also important to know the extent to which the research scholars are aware of the electronic information resources and services, and also to what extent they use them for their research work.

The investigator wishes to conduct a study about the user pattern of electronic information resources in the University libraries in Tamil Nadu, because the present study focuses on the seamless usability of the electronic information resources and its growth, development and maximum use of these most modern electronic information are taken for a clear cut study and assessment. Today□s Research scholars may be more sophisticated than their predecessors in using some of the basic functions of many systems, however many are clearly unaware of the limitations and use of many electronic sources of information. Research scholars and faculty members entering research appear to have raised expectations about the capabilities of electronic resources. So this type of study is significant and important in the new IT Era.

1.7 OBJECTIVES OF THE STUDY

The specific objectives of the study are as follows:

1. To identify the purpose and frequency of visit to the library.
2. To know the type of electronic resources accessed by users
3. To study the time spent on access of the electronic resources by the users
4. To study the awareness on E-Resources by faculty and research scholars.
5. To determine the level of satisfaction of the users on E-Resources.
6. To study the importance of using E-resources.
7. To identify the problems while accessing electronic information services.

1.8 HYPOTHESIS OF THE STUDY

There is no significance difference between the respondent categories of research scholars and faculties towards the purpose of using library

There is no significance difference between the respondent categories of research scholars and faculties towards the extent of satisfaction with electronic information resources.

There is no significance difference between the respondent categories of research scholars and faculties towards the problems facing in accessing electronic information services.

1.9 METHODOLOGY

The study attempts to report the findings of Information Literacy through Madras University Library System, Tamil Nadu. The primary data were collected through questionnaire. Both Primary and Secondary data were collected and analyzed. Since this is a time-bound study at the state level, necessary mathematical and statistical methods and tools are applied.

1.9.1 Sampling Technique

There are 250 questionnaires were distributed to the Research Scholars in three major disciplines such as Humanities, Social Science and Science at Madras University in Tamil based on convenience sampling method the questionnaires were distributed to the available research scholars at the time of data collection. Among the 275 research scholars 200 questionnaires were retrieved, and used for the analysis.

1.9.2 Study Period

The data was collected from the research scholars at Madras University in Tamil Nadu during January 2018 to June 2018 for the present study.

1.9.3 Statistical Techniques Used

Both Primary and Secondary data were collected and analyzed. The Entire Statistical analysis was done using "Statistical Package for Social Sciences" (SPSS) software. SPSS provides a statistical analysis and data management system in a graphical environment.

1.10 LIMITATION OF THE STUDY

This study is limited to only Madras University Library, Tamil Nadu.

1.11 ORGANIZATION OF THE THESIS

The study has been divided into five chapters including the introduction and conclusion. At the end of the work, bibliography & references and annexure are also included.

Chapter 1: Introduction: In the first chapter the research design of the study area is included with elaboration; the subject of the study has been introduced illustratively including the field of study, objective of the study, methodology, hypotheses, problem of the study, need and significance of the study, scope and limitations of the study and organization of the report.

Chapter 2: Review of literatures: This chapter gives an appraisal of the relevant literature with the study. Literature search includes the available updated development of various concepts. Both micro and macro literature appeared in printed and digital form are included here. The documents have been arranged first according to the concept and then arranged within each concept chronologically.

Chapter 3: Madras University in Chennai: This chapter is a description of the area of the study

Chapter 4: This chapter confined to analysis and interpretation of data.

Chapter 5: This chapter covers the summary of the findings of analysis, tenability of hypothesis, suggestions for the improvement and further research. At the end of the work Bibliography was included followed by annexure.

REFERENCES

1. Barry M Leiner, "The Scope of the Digital Library, January 16, 1998
2. Marchionini, G. Research and development in digital libraries, Allen Kent (Ed).
3. Encyclopedia of Library and Information Science, 63: 259-279, 1998
4. ICDL 2004. www.teriin.org/events/icdl/background.htm
5. Bruce Perens, "The Debian Free Software Guidelines" June 1997. www.opensource.org/docs/definition.php
6. Frost, et al., "Some Economic and Legal aspects of Open Source Software, May 7th 2005".

Chapter - II

Review of Literature

2.1 PREAMBLE

This chapter deals with the analysis of the past trends in research output in academic research. It enables one to identify the area of research concentration in any particular branch of science. This analysis helps to explore the content and meaning of research done over the years. Research in the field of use and availability of electronic resources and this study of utilization and adoption of ICT"s is a new milestone in the developing field of Library and Information Science. The previous research finding gives a clue_about the potential areas of research and missing links in the process of research in information science. The Review of Literature enables the researchers to identify the gaps in the previous studies and helps them to undertake this study in a systematic manner

The review of literature are categorized as follows

1. Studies based on E-Resources and Services in Academic Environment
2. Studies based on E-Resources and Services in Digital Environment
3. Studies based on E-Resources and Services in Web Environment
4. Studies based on E-Resources and Services among Academicians

2.2 STUDIES BASED ON E-RESOURCES AND SERVICES IN ACADEMIC ENVIRONMENT

Ellis, et al. (1993)[1] attempted to propose and describe the characteristics of a general model of information-seeking behaviors based on studies of the information-seeking patterns of social scientists, research physicists and chemists, engineers and research scientists in an industrial firm. Ellis's elaboration model describes the features of information-seeking activities as generic.

Dhanavandan and Esmail (2000)[2] in their study found that 40.82% of the staff used the digital resources to collect general knowledge and 40.81% learned digital resources through training provided by the librarian.

Thelwall and et al. (2001)[3] give a new approach for specific information where a website that enables a user to search the individual websites. The Google, AltaVista and Hotpot offer the facility to integrate additional information into a more advanced request

Kebede (2002)[4] identifies four barriers to the effective provision of electronic resources in those libraries, namely: lack of strategic planning: lack of adequate or reliable funding; lack of use of Internet to provide information services to users and a lack of consistent training for users in new ICT services.

Pandey and Mehta (2002)[5] conducted a study on awareness of educational technologies in open learning system by target groups. A majority of the respondents were from an active age group, unmarried, unemployed, and belonged to nuclear families and were having services as the main family occupation.

Ajuwon (2003)[6] carried out a study of uptake of ICTs by health science students at the University College Hospital, Ibadan. This study found that 57% of students sampled could not use a computer, that the use of the database was poor, due to lack of awareness, lack of access to computers, insufficient training and the high cost of provision

Appleton(2004)[7] concluded that e-books have a place in health library and information resources, but further development of e-books and e-book collections is required and subsequent investigation into their most effective use.

Doney et al. (2005)[8] found that most respondents reported used the Internet (81%), but fewer (44%) reported used databases. GPs and PNs were significantly more likely to report using databases than PMs ($P < 0.01$). Lack of training was the most reported barrier to using the Internet (67%) and databases (52%). Overall, 52% of respondents reported wanting Internet training, 64% wanted database training.

Haneefa (2007)[9] concluded the email service was used by the largest percentage of the users. WWW was being used by 60 per cent of the library users. A good number of users were not satisfied with the application of ICT in the libraries and indicated „inadequate ICT infrastructure" as their reason for dissatisfaction. Users proposed a variety of measures of formal orientation and training in ICT based resources and services.

Sujatha and Mudhol (2008)[10] surveyed that the 88.9% users used electronic information sources for thesis/ project work and 87.6% used for ongoing research work. A majority of the respondents comprising 79.0% browsed the Internet for subject specific information websites and 75.3% for international/ regional institution websites. E-mail was the most frequently used electronic information source followed by web resources.

Swain and Panda, (2009)[11] concluded that the internet-based e-resources are being well used compared with CD-ROM databases. Premier web search options like Google and Yahoo! are the most frequently used search engines, while the other searching options are less used. Availability of some key online databases are exclusively confined to only a selected few B-School libraries of the State.

Agarwal and Dave (2009)[12] studied the use of internet by the scientists and research fellows of Central Arid Zone Research Institute, Jodhpur (Rajasthan) through a questionnaire survey. The study revealed that respondents accessed Google search frequently (100%) followed by Yahoo (85.29%) and 97.06% of respondents use the internet for education and research.

Sami and Iffat (2009)[13] in paper "Use of Electronic Information Services in Research Libraries" discussed the use of electronic information services in research libraries and attempted to reveal the impact of users" background and awareness about services on the use of electronic information services

Trivedi and Joshi (2009)[14] revealed that there was more use of print journals (54.63%) as compared to e-journals. Authors concluded that there is a growing use of e- journals over print journals

Thanuskodi (2010)[15] revealed that the Internet and CD-ROM were the most frequently used IT-based sources and facilities. Results of the present study show that majority of students do not have a personal computer or laptop. The study reveals that an majority of the respondents (57.28%) feel that the internet and electronic resources cannot replace the print resources.

Sasireka et al. (2011)[16] found that out of 205 institutions, 54.2% of the institutions were established after the year 2000. Only 45.8% of the institutions were established before the year 2000. Online journals (88.3%) and e-journals (77.1%) are most preferred digital resources (table-2). Most of the institutions are preferring consortium for subscribing e-resources.

Selvi and Dhanavandan (2012)[17] conducted a study and found that respondents were using internet every day, or once in a week and 82.22% of the respondents used Google and 70.22%yahoo search engine.

Nwagwu et al. (2013)[18] revealed breweries need to recognize and improve the availability and accessibility of the Internet in the workplace; they also need to promote social Web technologies such as blogs, wikis, podcasts and other social networking services to increase social interaction with colleagues and other persons. These strategies ultimately lead to better job performance, facilitate innovation, and encourage economic growth in the breweries.

Coates (2013)[19] stated that the confusion regarding the nature of e-books slows adoption. While participants were exploring ways to incorporate e-books into their norms, values, and research practices, they are unlikely to rely solely on e-books as primary sources.

Joanna Duy, Liwen Vaughan (2015)[20] revealed Vendor-provided electronic resource usage statistics are not currently standardized across vendors. This study investigated the feasibility of using locally collected data to check the reliability of vendor-provided data. The study finds that the two types of data correlate well in terms of use patterns, but that actual usage values differ for many products.

Debbie Malone, Carol Videon (2015)[21] found between instruction in the use of electronic resources and increased usage of these sources. Three institutions whose students used a number of electronic resources did provide useful teaching models for further study.

Maharg, Paul and Paliwala, Abdul (2015)[22] studied the relationship between electronic learning resources and the pedagogy of legal education. Aimed at teachers of both undergraduate and postgraduates in law explores how the conceptual framework of resource-based learning (RBL) can be adapted to support the new technologies of ICT.

Lohar and Roopashree (2015)[23] found a collected data that covers the use of electronic resources and how the electronic resources are improving the academic carrier of the faculty and also the problems faced in using the electronic resources. Concluded that the main intention of the use of electronic resources has been the academic interest of the users.

Barbara Bren, Beth Hillemann and Victoria Topp (2015)[25] reported that there were no differences related to gender. The results of this study reinforce current views on the effectiveness of active learning and lend support to the continued development of multiple-workstation classrooms.

Holleman, Curt (2016)[26] analyzed that they are price, consortial discounts, search engines, accessibility, content, current usefulness, and lasting benefit. It is quite another matter to see the interplay of these and other concerns as vendors attempt to create appeal for their products and services. Some offerings provide such good deals that they can hardly be refused. Some are initially wonderful opportunities but give pause over time. Some are excellent agreements for some participants and mediocre ones for others.

Celeste Feather (2016)[27] reported that the communications in the workflow of electronic resources (e-resources) acquisitions and management are complex and numerou and revealed the underlying structure of the communication network and areas that could be improved in terms of efficiency and effectiveness. E-resources management must be responsive to the high expectations of users and other library staff. Efficient management of the related communications network increases the likelihood of a productive and successful operation.

2.3 STUDIES BASED ON E-RESOURCES AND SERVICES IN DIGITAL ENVIRONMENT

Francis" (2005)[28] examination of the information seeking behavior at the University of West Indies revealed that a preference for journal articles in electronic format over print articles, which demonstrated that social scientists "have embraced electronic resources.

Weiler (2005)[29] explores the search for possible motivating factors behind students" dependence on television and the internet for their information needs. The research indicates that only a very small percentage of the general population prefers to learn by reading.

Asemi and Riyahiniya(2007)[30] found that 70 percent of students were aware of digital resources, but only 69 percent of them have used them; 62 percent were aware of offline databases, whereas only about 19 percent used them through the Central Library LAN network. About 70 percent were aware of online databases, accessible via the Central Library web site, and about 53 percent of respondents have used them.

Shuling (2007)[31] indicated the homepage had become the important window to give publicity and serve the outside. It also shows the use of reading on screen, non- familiarizing the structure and the retrieval method. Showing that the net condition and computer facilities are the main external factors

Kumar and Kumar (2008)[32] highlighted six reasons for using electronic information sources. Users in the study use electronic sources in support of their study (70%) and teaching (59%). One-third of respondents used the sources for project work.

Kindilchie and Samarraie(2008)[33] revealed that the databases and e-resources provided by the University were not well utilized by the entire faculty, although they are making a difference in the work of some of the University's faculty members. The University Library needs to undertake a review to identify other databases that may be needed, to promote the use of e- resources in teaching and learning, and to develop the search skills of faculty and students

Vakkari(2008)[34] identified that the first one is improved accessibility and availability of literature, and the second is more directly related to the content and quality of scholarly work. The perceived improved access is positively associated with the number of international publications produced, among doctoral students in particular.

Atakan et al. (2008)[35] measured the effectiveness of digital library. Investigators compared the results of two surveys carried out in 2002 and 2005. Results revealed that many faculty members were using electronic databases. Web of Science, Science Direct and Ebsco were the most preferred databases.

Lakan (2008)[36] revealed that power outrages, slow Internet service and insufficient computers were the factors affecting the provision and use of electronic information resources.

Thangaraja et al. (2008)[37] attempted a study on the utilization behavior of online journals by the students of Tamil Nadu Agricultural University (TNAU). The results showed that a majority (95%) of the students were aware about the availability of the online journals in their respective subjects

Gowda and Shivalingaiah (2009)[38] found that nearly 58.12% respondents indicated their preference towards print resources over electronic resources. The preferences of users depended upon the nature of resources available in libraries, while the usage of electronic resources depended upon the type of IT infrastructure available in libraries

Joteen Singh et al. (2009)[39] executed a study under the title "Use of Internet Based E-Resources at Manipur University: A Survey" to examine the use of electronic information focusing on the Internet services by post graduate students, research scholars, teachers and non-teaching staff members. Users were accessing the Internet mainly to download the information from web based resources and web sites.

Okello-Obura (2010)[40] found that the respondents have positive attitudes towards e- resources utilization and they indicated varied problems faced in e-resources utilization and proposed different strategies to improve on e-resources in the University. Among the problems identified are: slow Internet connectivity, inadequate networked computers, lack of access to low cost printers in the library, using advanced search strategies of most databases and lack of awareness of most of the e-resources. Boumarafi (2010)[41] showed that the evidence of use of e-resources, but they did not explicitly establish that some of the major problems mentioned by participants did hinder the information searches of the respondents.

Govindaraju (2010)[42] found by his survey that the use of electronic resources is found to be significant among the users of the Andhra University. This is quite natural and expected in the present day implementation environment. Some of the resources such as e-book, encyclopedias, e-dictionaries are less used. He also felt that to further maximize use of the e-resources, wide publicity and imparting trainings were found to be necessary.

Natarajan, et al (2010)[43] revealed that despite the availability of a wide range of e- **resources** the frequency of their use was low. The reasons identified for this are lack of time, lack of awareness, lack of subject coverage; and slow downloading

Ollé et al. (2010)[44] found that the participants confirmed the increase in the amount of journal reading among academics, due to the increase in the number of **electronic** journals available and the improvement in the tools for locating and accessing this information, especially off-campus access. Librarians were well aware of the importance of internet search engines as the first information source for academics. They regretted having failed to provide users with a single interface to all information products..

Mc Clanahanand et al. (2010)[45] found that their use of e-journals, Internet resources, and email communication has become indispensable to their scholarly activities, and has profoundly changed the way they interact with information, create new knowledge, and relate to their university library. For these faculty members, e-journals have supplanted the print format, because of their convenience, flexibility of work location, advantages for interdisciplinary access, and ease of storage.

Madhusudhan (2010)[46] studied and found that e-resources can be good substitutes for conventional resources, if the access is fast, and more computer terminals are installed to provide fast access to e-resources. Google is the most widely used search engine for locating information electronically.

Sheeja (2010)[47] found the information seeking behavior of science and social science research scholars, including service effectiveness, satisfaction level on different type of sources and various methods adopted by the scholars for keeping up to date.

Sudhier and Seethalekshmi (2011)[48] found from the analysis of the data given by the respondents that most of the Arts department are not giving enough facility for the use of e-resources.

Thanuskodi (2011)[49] revealed that the M.Phil students undertake more searching of e-resources followed by postgraduate students and PhD Scholars the least. It also indicates that users are aware of the e-resources and their various types but recommend the improvement in the access facilities.

Halima Sadia Egberongbe (2011)[50] showed that the uses of e-resources are very common among the lecturers and research scholars of the University of Lagos. It also showed that a majority of teachers and research scholars are dependent on e-resources to obtain desired and relevant information.

Zha, Xianjin et al. (2012)[51] explored the exact nature of patrons" usage behavior as well as their perception of ease of use and usefulness of print and electronic resources in the context of Chinese university libraries. The study found that users are undergoing a transfer of usage behavior from print resources to electronic resources.

Pradeep Mittal and Bonu Bala (2013)[54] concluded that there is non-availability of training programmes in universities regarding the need and use of e-Resources. Most of the research scholars are not aware of these e-resources which can be very helpful for their research and can make it very fast and effective.

Nzivo, Charles N and Chuanfu, Chen(2013)[56] studied that the Chinese academic libraries are considerably well perceived by international students. It also indicates that they have differing purposes, and differing needs for library services and information resources.

Asunka, Stephen (2013)[58] revealed that students generally have a fair idea of what e-books are, and also acknowledge the potential benefits of e-textbooks. Factors influencing these **perceptions** are discussed within the broader context of barriers to effective e-textbook adoption and usage in developing countries.

Anaraki and Babalhavaeji(2013)[59] found that the awareness and utilization level of students in the three universities were lower than the average and those who are not aware of the existence of the IDL portal used general search engines to meet their information needs.

Seema Vasishta (2014)[60] showed that majority of research scholars and faculty are aware of benefits of e-resources and their use is very common among them. However a few constrains are also reported by the users.

Dhanavandan (2014)[61] assessed and found that respondents stated that the arrangement of reading material in the library is easy to access and 34.29% respondents are satisfied.

Alia Arshad, Kanwal Ameen (2015)[62] revealed that the university library website was heavily used on campus by users; however, it was used by a significant number of clients residing off-campus. Free scholarly journals, resources downloaded, e-journals, e- books and donated personal collections were among the top most used resources and services. However, free scholarly journals were used more than subscribed e-journals accessible on-campus only.

2.4 STUDIES BASED ON E-RESOURCES AND SERVICES IN ICT ENVIRONMENT

Adikata and Anwar (2006)[126] a study focusing the role of the faculty in making the students to use the library was conducted in International Islamic University,

Malaysia. The study revealed that teachers were instrumental in promoting the usage of libraries by the students. However the student library use skills were found unsatisfactory and therefore librarians were expected to provide a comfortable academic environment in thelibrary.

Albornoz et al. (2007)[127] have conducted a study on information searching and library use by researchers in humanities at the University of Argentina. The main objective of this research was to identify the patterns of behaviour of the researches at the Faculty of Humanities and Education and their use of information resources within university libraries.

Rafiq and Ameen (2009)[128] investigated information-seeking behaviour and satisfaction level of instructors at the National Textile University, Pakistan. Questionnaires were distributed for collecting data among 35 teachers and 34 questionnaires were received. The study brought out the information seeking behaviour of a community that was engaged in research and development in the field of textiles, which was the backbone of Pakistan's commerce and industry.

Patil and Parameshwar (2009)[129] undertook a study to find out the information needs and usage of electronic resources by the faculty members and research scholars . The study showed that the information needs were satisfied using the electronic resources available in the University Library. It was recommended to include more number of journals in the UGC-Infonet Consortium.

Gowda and Shivalingaiah (2009)[130] showed that for a majority of the humanities and social science respondents, the facilities and services available in the university libraries were poor whereas the science respondents opined that the facilities and services were moderately good. Results showed that there were significant differences in the satisfaction level of facilities and services among the research scholars of various disciplines.

Devendra Kumar (2010)[131] conducted an analytical study of information seeking behaviour among Agricultural Scientists in Sardar Vallabhbhai Patel University of Agriculture and Technology. This study employed a structured questionnaire which was distributed among scientists in various teaching and research departments of the Sardar Vallabhbhai Patel University of Agriculture and Technology.

Khan, Bhatti and Khan (2011)[132] analyzed from the data collected through questionnaires from Principals, Vice Principals and teachers of the Government colleges selected in Bahawalpur. It was revealed that information technology have affected the users information seeking behaviour and made the information process easier.

Drachen and Asger (2011)[133] conducted a study that consisted of a survey and interviews focused on the PhD students at three universities. The study recommended to provide specialized and to the point course at the beginning of the Ph.D programme libraries should teach Ph.D scholars most efficient methods for searching early in their studies, to provide dedicated PhD work space etc.

Chopkar and Khaparde (2012)[134] revealed that 120 (89%) research scholars visit the library. Most of the research scholars visit the library twice a week. The first preferred sources of the information are books followed by newspaper and journals. The paper also discusses knowledge and use of electronic sources, problem in finding information and satisfaction of research scholars for library resources and services.

Khan (2012)[135] carried out a study on the Use of Information Sources by Faculty Members, Research scholars and Students of the Faculty of Commerce, AMU, Aligarh: A Survey. In this research, it was discovered that most of the faculty members as well as research scholars consulted Emeraldinsight.com and Science Direct.com for accessing their required information.

Bitso and Fourie (2012)[136] revealed work roles, associated tasks, information needs and information seeking patterns. An information service model for Lesotho secondary level geography teachers is proposed.

Mostofa (2013)[137] conducted a study and the results show that half the faculty members are consulted with expert as communication channel when they have a question asked. More than 50 % of faculty members depend on the library resources for teaching purpose.

Fasola and Olabode (2014)[138] surveyed among the students of ajayi crowther university, oyo, Nigeria, how they seek information and this is what has prompted this study. It was discovered that a majority of the students (66%) sought information for academic purposes and the library (62.8%) was their preferred place of searching and using information resources.

Acheampong and Dzandu (2015)[139] conducted a study among crop research scientists seek information in terms of what information sources they use, type, when, where and the purpose for which they use information as well as how they search for information. The researcher employed a questionnaire to collect data from 73 scientists of the Institute, while three Library staff were interviewed on their approach to meeting the information needs of the scientists. They used information centres/libraries for research purposes and prefer to use journals articles but preferred electronic format to other types of materials.

Ngozi, Uche and Ejiro (2015)[140] investigated the information seeking behaviour of faculty members of the Federal University of Petroleum Resources (FUPRE). The research finding shows that the respondents use books, as their preferred source of information. They suggested to improve the internet facilities to assist faculties in their research.

Mahapatra (2017)[141] has attempted to collect information related to the preferences on electronic information resources, types of e-resources used, the use of statistical information in e-format, online databases in social science and satisfaction on the use of e-resources. The information has been analysed in the light of data collected from 90 social scientists from research institutes and universities in the city of Bhubaneswar.

Gopinath (2017)[142] The study intends to assess the perception and use of electronic information resources on the quality of education and research among the academic community in the Mahatma Gandhi University. The study applied a standard survey. On the basis of the results, some suggestions have been put forward for improving the use of electronic information resources among the academic community in the Mahatma Gandhi University.

REFERENCES

1. Ellis, D., Cox, D., and Hall, K., (1993). A Comparison of the Information-Seeking Patterns of Researchers in the Physical and Social Sciences. *Journal of Documentation*, 49(4), 356-369.
2. Dhanavandan,S., Mohamed Esmail,S. andSivaraj,S.,(2000). Use of digital resources among Engineering Professionals in Cuddalore District, Tamil Nadu, *Indian Journal of Information Science and Services*, 3(1), 36-1.
3. Thelwall, Mike et al. (2001). Custom interfaces for advanced queries in search engines. *ASLIB Proceedings,* 53(10), 413-422.
4. Kebede, G. (2002). The changing information needs of users in electronic environments. *The Electronic Library*, 20(1), 19 - 21.
5. Pandey,R., and Mehta,S., (2002). Awareness if Educational Technologies in open learning system by Target group. *Indian Journal of Social Research*, 43(3), 183-189.
6. Ajuwon G. A. (2003). Computer and Internet Use by First Year Clinical and Nursing students in a Nigerian Teaching Hospital. *BMC Medical Informatics and Decision Making*, 3(10).
7. Appleton, L. (2004). The use of electronic books in midwifery education: the student perspective. *Health information and libraries journal,* 21(4), 245-252.
8. Doney, L., Barlow, H., West, J.,(2005). Use of libraries and electronic information resources by primary care staff: outcomes from a survey. *Health information and libraries journal*, 22(3), 182-188.
9. Haneefa, K. (2007). Use of ICT based resources and services in special libraries in Kerala. *Annals of Library and Information Studies,* 54(1), 23-31.
10. Sujatha, H.R., and Mudhol, Mahesh V., (2008). Use of Electronic Information Sources at the College of Fisheries, Mangalore, India. *Annals of Library and Information Studies*, 55(3), 234-245.
11. Swain, D.K., Panda, K.C., (2009). Use of electronic resources in business school libraries of an Indian state: A study of librarians' opinion. *Electronic Library*, 27(1), 74-85.

12. Agarwal, Umesh Kumar and Dave, Rajesh Kumar. (2009). Use of Internet by the Scientists of CAZRI: A survey. *Indian Journal of Library and Information Science*, 3(1), 15-21.

13. Sami, Lalitha K; and Iffat, Rabia. (2009). Use of Electronic Information Services in Research Libraries. *SRELS Journal of Information Management*, 46(1), 43-50.

14. Trivedi, Mayank., and Joshi, Anuradha., (2009). Usage of Electronic Journals (E- Journals) Versus Print Journals by Healthcare Professionals in H M Patel Centre for Medical Care and Education (HMPCME). *Journal of Health Informatics in Developing Countries*, 3(1), 13-16.

15. Thanuskodi, S. (2010). Use of internet and electronic resources for agricultural science information: A case study. *Social Sciences*, 5(4), 364-367.

16. Sasireka, G., Balamurugan, S., Gnanasekaran, D., (2011). Use of E-resources in digital environment among engineering institutions in Tamil Nadu (India): An empirical study. *European Journal of Scientific Research*, 60(3), 326-333.

17. Selvi,M., and Dhanavandan,S., (2012). A Study on the Utilization of Interne Resources and Tools by Basic Medical Science Professionals. *Journal of Emerging Trends in Computing and Information Sciences*, 3(12), 1634-1637.

18. Nwagwu, W.E.ab.,Segilola, M.O.a., (2013). Information needs information sources and information seeking behaviours of engineers in breweries in Nigeria. *Turkish online journal of education*, 63(2), 159-178.

19. Coates, H. (2013).Academic historians in Canada report both positive and negative attitudes towards E-books for teaching and research. *Evidence Based Library and Information Practice*. 8(4), 148-150.

20. Joanna Duy, Liwen Vaughan (2015), Usage data for electronic resources: A comparison between locally collected and vendor-provided statistics, *The Journal of Academic Librarianship*, 29 (1),16-22.

21. Debbie Malone, Carol Videon (2015). Assessing undergraduate use of electronic resources: A quantitative analysis of works cited, *Research Strategies*, 15(3),151-158.

22. Maharg, Paul and Paliwala, Abdul (2015). Negotiating the learning process with electronic resources. In: Effective learning and teaching in law. Kogan Page, London, UK, 81-104.

23. Lohar M.S and Roopashree T.N (2015). Use of Electronic Resources by Faculty Members in B.I.E.T, Davanagere: A Survey, *Science*, 43 (1), 201.

24. Jonathan P. Egle, David M. Smeenge, Kamal M. Kassem, Vijay K. Mittal (2015).

25. The Internet School of Medicine: Use of Electronic Resources by Medical Trainees and the Reliability of those Resources, *Journal of Surgical Education*, 72(2), 316– 320.

26. Barbara Bren, Beth Hillemann, Victoria Topp (2015), Effectiveness of hands-on instruction of electronic resources, *Research Strategies*, 16(1), 41-51.

27. Holleman, Curt (2016). Electronic Resources: Are Basic Criteria for the Selection of Materials Changing?, *Library Trends Spring*,694.

28. Celeste Feather (2016). Electronic Resources Communications Management: A Strategy for Success, the Electronic Resources and Libraries Conference, Georgia Institute of Technology

29. Francis, H., (2005). The information seeking behaviour of Social Science Faculty at the University of the West Indies, St. Augustine Campus, *Journal of Academic Librarianship*, 31(5), 67-72.

30. Weiler, Angela. (2005). Information-Seeking Behavior in Generation Y Students: Motivation, Critical Thinking, and Learning Theory, *Journal of Academic Librarianship*, 31(1), 46-53.

31. Asemi, A., and Riyahiniya, N., (2007). Awareness and use of digital resources in the libraries of Isfahan University of Medical Sciences – Iran, *Electronic Library*, 25(3),316-327.

32. Shuling, W., (2007).Investigation and analysis of current use of electronic resources in university libraries, *Library Management*, 28(1-2), 72-88.

33. Kumar, G.T., and Kumar, B.T. Sampath., (2008). Use of electronic information sources by the academic community: A comparative study. *6th International CALIBER-2008*, Allahabad: University of Allahabad, 684-692.

34. Kindilchie, A.I., and Samarraie, I.F., (2008). Interaction and impact of electronic information resources on Qatar university faculty, *Libri*, 58(4), 281-293.

35. Vakkari, P.,(2008).Perceived influence of the use of electronic information resources on scholarly work and publication productivity, *Journal of the American Society for Information Science and Technology*, 59(4), 602-612.

36. Atakan, Cemal. Atilgan, Dogan., Bayram, Ozlem., and Arslantekin, Sacit., (2008). An Evaluation of the Second Survey on Electronic Databases Usage at Ankara University Digital Library, *The Electronic Library*, 26(2), 249-259.

37. Lakan, E.L., (2008). Availability and Utilization of Electronic Information Databases by Staff of the Agricultural Complex, Ahmadu Bello University, Zaria, *Samaru Journal of Information Studies*. 8(1), 44-53.

38. Thangaraja,K., Karthikeyan,C.,Asokan, M., and Rajasekaran, R., (2008). Study on the Utilization behaviour of online journals by the students of TNAU, *The Madras Agricultural Journal,* 95(1-6), 114-119.

39. Gowda, Vasappa., and Shivalingaiah, D., (2009). Attitude of Research Scholars towards Usage of Electronic Information Resources: A Survey of University Libraries in Karnataka, Annals *of Library and Information Studies*, 56(3), 184-191.

40. Joteen Singh, R.K., Devi, Th. Madhuri., and Raychaudhury, Arup., (2009). Use of Internet Based E-Resources at Manipur University: A Survey, Annals *of Library and Information Studies*, 56(1), 52-57.

41. Okello-Obura.C., (2010). Assessment of the problems LIS Postgraduate students face in accessing e-resources in Makerere University, Uganda, *Collection Building, 29* (3), 98-105

42. Boumarafi, B., (2010). Electronic resources at the university of Sharjah medical library: An investigation of students' information-seeking behavior, *Medical Reference Services Quarterly, 29* (4), 349-362.

43. Govindaraju, Nemani., (2010). Use and user awareness of E-resources in Andhra University library: A study, PEARL*: A Journal of Library and Information Science*, 4(3), 183-188.

44. Natarajan, K., Suresh, B., Sivaraman.P, and Sevukan, R., (2010). Use and user perception of electronic resources in Annamalai University: A case study, *Annals of Library & Information Studies*, 57(1), 59-64.

45. Ollé, Candela., and Borrego, Ángel., (2010). Librarians' perceptions on the use of electronic resources at Catalan academic libraries: results of a focus group, *New Library World*, 111(1/2), 46-54.

46. McClanahan, K., Wu, L., Tenopir, C., and King, D.W., (2010). Embracing change: Perceptions of e-journals by faculty members, *Learned Publishing*, 23(3), 209-223.

47. Madhusudhan, M., (2010).Use of electronic resources by research scholars of Kurukshetra University, *Electronic*, 28(4), 492-506.

48. Sheeja, N., (2010). Science vs Social Science: A study of information-seeking behavior and user perceptions of academic researchers, *Library Review*, .59(7), 522- 530.

49. Sudhier.K.G., and Seethalekshmi, K P., (2011). Use of E-resources by the students and Researchers of Faculty of Arts, University of Kerala, International *Journal of Information Dissemination and Technology*, 1(3),

50. Thanuskodi, S., (2011). Usage of electronic resources at Dr T.P.M. Library, Madurai Kamaraj University: A case study, *DESIDOC Journal of Library & Information Technology*, 31(6), 437-445.

51. Halima SadiaEgberongbe (2011).The use and Impact of Electronic Resources at the University of Lagos, *Library Philosophy and Practice*,

52. Zha, Xianjin., Li, Jing., and Yan, Yalan.,(2012). Understanding Usage Transfer from Print Resources to Electronic Resources: A Survey of Users of Chinese University libraries, *Serials Review*, 38 (2), 93-98.

53. Santosh A Navalur, Balasubramani.R, Ashok Kumar.P, (2012). Usage of E-resources by Faculty, Research Scholars and PG Students of Bharathidasan University: A Study, *Journal of Advances in Library and Information Science*, 1(4), 165-172.

54. UmmeHabiba and Salma Chowdhury (2012), Use of electronic resources and its impact: A Study on Dhaka University Library users, *The eastern Librarian*, 23(1), 74-90.

55. Pradeep Mittal and BonuBala (2013).Use of e-Resources in Universities, *International Journal of Innovative Research in Computer and Communication Engineering*, 1(6)

56. Yan, Yalan., Zha. Xianjin.,and Xiao, Zhiliang., (2013). Exploring Users' Perceptions of Conventional and Unconventional electronic resources, *Serials Review*, 39(2), 105- 113.

57. Nzivo, Charles N., and Chafe, Chen., (2013). International Students' Perception of Library Services and information resources in Chinese academic libraries, *Journal of Academic Librarianship*, 39(2), 129-137.

58. Amjad, A.., Ahmed, S.., Naeem, S.B., (2013). Use of Electronic Information Resources among Research Scholars in the Islamia University of Bahawalpur, *Pakistan New Review of Academic Librarianship*, 19(3), 316-328.

59. Asunka, Stephen. (2013). The viability of e-textbooks in developing countries: Ghanaian University students perceptions, *Open Learning*, 28(1), 36-50.

60. Anaraki, L.N. , Babalhavaeji, F., (2013). Investigating the awareness and ability of medical students in using electronic resources of the integrated digital library portal of Iran: A comparative study. *Electronic Library*, 31(1), 70-83.

61. SeemaVasishta., (2014). Use pattern of E-resources by research scholars and faculty: A survey of technical university libraries in North India,*International Journal of Library and Information Studies*, 4 (1).

62. Dhanavandan, S., (2014). Knowledge and utilization of e-resources and services in Annamalai University Library, *International Journal of Information Dissemination and Technology.*

63. Alia Arshad, Kanwal Ameen , (2015). Usage patterns of Punjab University Library website: a transactional log analysis study, *The Electronic Library*, 33(1), 65 – 74.

64. Adikata1, A.A. and Anwar, M.A. (2006). OP. cit Albornoz, S. et al. (2007), "Information searching and library use by researchers in the humanities at the University of Argentina", *Revista Interamericana de Bibliotecologia*, 30(1),73-92.

65. Rafiq, M. and Ameen, K. (2009), "Information seeking behaviour and user satisfaction of university instructors: A case study", Library Philosophy and Practice, Retrieved on 26.12.2015 from www.webpages.uidaho.edu /~mbolin/rafiq-ameen.htm.

66. Patil, D.B. and Parameshwar, S. (2009) OP. cit.

67. Gowda, V and Shivalingaiah, D. (2009), "Attitude of research scholars towards usage of electronic information resources: A survey of university libraries in Karnataka", *Annals of Library and Information studies*, 56(3), 184-191.

68. Devendra Kumar (2010). OP. cit.

69. Khan, S.A. and Bhatti, R and Khan, A.A. (2011), "User of ICT by Students: A Survey of Faculty of Education at IUB", Library Philosophy and Practice (e-journal). Retrieved on 16.01.2016 from http://digitalcommons.unl.edu/ libphilprac/677/

70. Drachen, T.M. and Asger, V.L. (2011), "Information behaviour and practices of PhD students Appendices". Retrieved on 16.01.2016 from https://halshs.

archives- ouvertes.fr/hprints-00599034/file/Information_behav iour_and_practices_of_PhD_stud ents.pdf

71. Chopkar, H.D. and Khaparde, V.S. (2012), "Information Seeking Behaviour of Research Scholars of Biological Science Department of Dr. Babasaheb Ambedkar Marathwada University: A Study", *International Journal of Information Dissemination and Technology* 1 (4), 241-243.

72. Khan, A.M. (2012), ".Use of Information Sources by Faculty Members, Research scholars and Students of the Faculty of Commerce, AMU, Aligarh: A Survey",Library Philosophy and Practice (e-journal). retrieved from: http://digitalcommons.unl.edu/cgi/viewcontent.cgi?article=1895&context=libphilprac.

73. Bitso, Constance and Ina, Fourie (2012) OP. cit.

74. Mostofa, S.M. (2013), A Study of Information Needs and Seeking Behaviour of Faculty Members of Darul Ihsan University in Bangladesh", Library Philosophy and Practice (e-journal). 983. http://digitalcommons.unl.edu/libphilprac/983.

75. Fasola, O.S. and Olabode, S.O. (2014), "Information seeking behaviour of students of ajayi crowther university, oyo, Nigeria", *Brazialian journal Information Science, Marília* 7(2).47-60.

76. Acheampong, L.D. and Dzandu, M (2015). OP. cit.

77. Ngozi, O V, Uche, EU and Ejiro, AK (2015), Information seeking behaviour of faculty members in a Nigerian University, *International Journal of Academic Research and Reflection,* 3(2), 95-102.

78. Gopinath, S. A. (2017). Perception and Use Electronic Information Resources Among the Academic Community: Acase study. International Journal of Information Dissemination and Technology, 7(2), 142-145.

79. Mahapatra, R. K. (2017). Use of e-Resources among Social Scientists in Selected Institutes in Bhubaneswar: Astudy. International Journal of Information Dissemination and Technology, 7(2), 102-106.

Chapter - III

Madras University Library System as a Learning Resource

INTRODUCTION

Madras University Library system consists of four campus Libraries to serve to the academic and research needs of the faculties and students of these campuses. Each library is an unique learning resource center because of the nature of their collections and the patrons they serve. The Madras University Library(MUL) consists of the collections of Humanities and social sciences, the Guindy campus library(GCL)y is considered as one of the best science library in the country located in the Guindy campus, Taramain Campus Library(TCL) is a library that caters to the needs of the academics of the Basic Medical sciences of the Taramani Campus and Marian Campus Library(MCL) is one of the leading library of the country for the Indian Languages and Literature.

ADVISORY COMMITTEE

The Syndicate Library Committee for Madras University Library System consists of the Syndicate members of the University: The following significant actions are taken by the committee for the improvement of all the four campus Libraries:

- Committee took all the steps to weed out the old ,out dated and soiled books from Madras University Library along with the refurbishing and restructuring of the Main Library building which is nearly century old.
- Policy decision for the four campus libraries for the digital initiations.
- Requesting and recommending the authority for proper manpower planning of the Madras University Library system.
- Monitoring the collection developments along with functional and service modalities of the system

Table 3.1 Details of Madras University Library System

		MUL	GCL	TCL	MCL
a.	Total area of the Library (in Sq. Mts.)	8000 sq.m	3080 sq.m	1100 sq.m	1100 sq.m
b.	Total Seating capacity	500 (Main Reading Hall) 200 (Current Periodical section) 30 (e-Resources Centre) 10 (OPAC Search) 10(Theses Unit)	100 in Main reading Hall and 130 in theses and current periodicals	130	110
c.	Working hours (on Working days, on Holidays, before examination, during examination, during vacation)	8.00 AM to 8.00 P.M on all the days of the year	On Working days - 8.00 am to 8.00 pm On Holidays - 10.00 am to 5.45 pm	9.00 AM - 6.00 PM On Holidays - 10.00 am to 5.45 pm	9.00 AM - 6.00 PM
d.	Layout of the library (individual reading carrels, lounge area for browsing and relaxed reading, IT zone for accessing e- resources)	Reading carrels for researchers OLIZ (Online information Zone) with conducive A/C environment for accessing e- resources is available	Reading carrels for researchersOLIZ (Online information Zone) with conducive A/C environment for accessing e-resources is available	Reading carrels for researchers OLIZ (Online information Zone)with conducive A/C environment for accessing e-resources is available	Reading carrels for researchers OLIZ (Online information Zone)with conducive A/C environment for accessing e-resources is available
e.	Clear and prominent display of floor plan; adequate sign boards; fire alarm; access to differently – abled users and mode of access to collection	Floor Stack Area(Ground Floor (Subject books), II & III (Back volumes of journals) Stack-IV- Ready Reference & RCE collections Stack -V with rare documents published before 1960 Signboards are provided at all stack areas & other user centered areas Talking Library section is created for visually impaired users	Adequate sign boards and display boards are made available for each section. The fire extinguishers and alarms are provided. Ramp is provided to facilitate differently abled users access to the resources of the library	Adequate sign boards and display boards are made available for each section. The fire extinguishers and alarms are provided. Ramp is provided to facilitate differently abled users access to the resources of the librar	Adequate sign boards and display boards are made available for each section. The fire extinguishers and alarms are provided. Ramp is provided to facilitate differently abled users access to the resources of the librar

Library holdings:

		MUL	GCL	TCL	MCL
a.	Print (Books, Journals, back volumes and theses)	Total No. of documents(Books, Journals, Theses, Reports & Manuscripts :3,28,470) Books: 2,94,234 Bound Volumes: 44,195 Theses(Ph.D. M.Phil & others): 7650 Reports: 2 Manuscripts : 3(Ramanujan original work) Pamphlets:7400	106750	Books : 15000 Journals: 45 Titles Back volumes : 8,396 Theses : 876	Books: 95,000 Theses(Ph.D. M.Phil & others): 2202
b.	Non Print (Micro-fiche, AV)	20 AV Educational Audio & video Cassettes	Not available	-	-

		MUL	GCL	TCL	MCL
c.	Electronic(E-books, e-journals)	e-Resources: 431 UGC NFONET DLC e-journals 5500 + UGC NFONET DLC e-database 16	e-journals 5 UGC NFONET DLC e-journals 5500 + UGC NFONET DLC e-database 16	UGC NFONET DLC e-journals 5500 + UGC NFONET DLC e-database 16	UGC NFONET DLC e-journals 5500 + UGC NFONET DLC e-database 16
d.	Special collections (e.g. Text books, reference books, standards & patents	3 Volumes of Ramanujan"s original Manuscripts Text books : 42000 Reference books : 10000	Textbooks and Reference books: 45000	Text Books : 11,642 Reference Books : 3,000	Text and reference books : 69786
e.	Book banks	SC/ST Book Bank Scheme : 2008-09 (1370 Textbooks)	Not available	Not available	Not available
f.	Question banks	600	250	100	100

Madras University library system is using the following tools for deploying to provide access to the collection:

		MUL	GCL	TCL	MCL
a.	OPAC	A comprehensive OPAC system is implemented for entire collections of the Library using SOUL 2.0	A comprehensive OPAC system is implemented for entire collections of the Library using SOUL 2.0	A comprehensive OPAC system is implemented for entire collections of the Library using SOUL and WIN/ISIS	A partial OPAC system is implemented for entire collections of the Library using SOUL 2.0
b.	Electronic Resource Management Package for e-journals	e-JOURNAL access is made with UGC-INFONET JCCC platform to access 5500+ journals of the consortium and it facilitates to search more than 10000 open access peer reviewed journals	e-JOURNAL access is made with UGC-INFONET JCCC platform to access 5500+ journals of the consortium and it facilitates to search more than 10000 open access peer reviewed journals	e-JOURNAL access is made with UGC- INFONET JCCC platform to access 5500+ journals of the consortium and it facilitates to search more than 10000 open access peer reviewed journals	e-JOURNAL access is made with UGC- INFONET JCCC platform to access 5500+ journals of the consortium and it facilitates to search more than 10000 open access peer reviewed journals
c.	Federated searching tools to search articles in multiple databases	JCCC platform	JCCC platform	JCCC platform	JCCC platform
d.	Library Website	Portal is available	www.unom.ac.in /library/Guindy	Library Web Portal	Library Web Portal
e.	In-house / remote access to e- publications	In house access to all the e-resources	- IP based intranet access to in-house journals and UGC- INFONET Digital Library Consortium. -Remote access to e-journals in process from Guindy Campus Library	-	-

ICT deployed in the library:

		MUL	GCL	TCL	MCL
a.	Library automation	Automated Document Management System (ADMS) Developed in all campus libraries Created Online Public Access Catalogue (OPAC) Established Online information Zone (OLIZ) Partial Digital Library Systems (DLS) are created e-Books, e- Databases, e- Journals are subscribed Collection building in print materials.	Automated Document Management System (ADMS) Developed in all campus libraries Created Online Public Access Catalogue (OPAC) Established Online information Zone (OLIZ) Partial Digital Library Systems (DLS) are created e-Books, e- Databases, e- Journals are subscribed Collection building in print materials.	Automated Document Management System (ADMS) Developed in all campus libraries Created Online Public Access Catalogue (OPAC) Established Online information Zone (OLIZ) Partial Digital Library Systems (DLS) are created e-Books, e- Databases, e- Journals are subscribed Collection building in print materials.	-
b.	Total number of computers for general access	45 numbers	15 No"s	49Nos	13 Nos
c.	Total numbers of printers for general access	2 printers	Nil	5 Nos	Nil
d.	Internet band width speed	100 mbps	100 mbps	100mbps	100 Mbps
e.	Institutional Repository	Nil	Nil	Nil	Nil
f.	Content management system for e-learning	Online tutorials for novice user	Nil	-	-
g.	Participation in resource sharing networks / Consortia (like INFLIBNET)	INFLIBNET - JCCC MALIBNET - Resources of all participating libraries in Chennai are shared	INFLIBNET - JCCC MALIBNET - Resources of all participating libraries in Chennai are shared	INFLIBNET - JCCC MALIBNET - Resources of all participating libraries in Chennai are shared	INFLIBN ET - JCCC MALIBN ET - Resources of all participati ng libraries in Chennai are shared

Details (Per month with regard to):

		MUL	GCL	TCL	MCL
a.	Average number of walk-ins	15000 per month	6000 per month	2250 per month	3000 per month
b.	Average number of books issued / returned	2000/per month	1000 Issued/retu rned	350/290	700 per month
c.	Ratio of library books to students enrolled	1500 : 1	500 : 1	105: 1	500 : 1
d.	Average number of books added during the last four years	5290 per year	2700 per year	1360/per year	3000 per year
e.	Average number of login to e-resources	5000 per month	6000 per month (Wi/fi connectivity)	3400 per month	1000 per month
f.	Average number of e –resources downloaded / printed	4500 per month	2500 per month	2700 per month	900 per month
g.	Number of IT (Information Technology) literacy trainings organized	24 per year	10 per year	05 per year	Nil

Specialized services provided by the library with regard to:

		MUL	GCL	TCL	MCL
a.	Manuscripts	Yes	Nil	Nil	Nil
b.	Reference service	Yes	Yes	Ready reference service, OPAC, Theses access service, SDI	Yes
c.	Reprography / Scanning service	Yes	Yes	Yes	Yes
d.	Inter-Library Loan service	Yes	Yes	MALIBNET	Yes
e.	Information Deployment and Notification	yes	Yes	Made through Library web portal	Yes
f.	OPAC	Yes	Yes	Databases of Books, journals and theses are available in OPAC	Yes
g.	Internet Access	Yes	Yes	Provided	Yes
h.	Downloads	Yes	Yes	Allowed	Yes
i.	Printouts	Yes	Nil	allowed to copy in Pen drives	Yes
j.	Reading list / bibliography compilation	Yes	Yes	Available	Yes
k.	In-house / remote access to e-resources	yes	Yes	Online subscribed journals and INFLIBNET journals	Yes
l.	User Orientation	yes	Yes	Provided	Nil
m.	Assistance in Searching Databases	Yes	Yes	Provided	Nil

Annual library budget and the amount spent for purchasing new books and journals:

MUL :	Books & Journals purchase	: 44,53,000/per year
GCL :	Books & Journals purchase	: 48,12,768
TCL :	Allocated amount	: 40,00,000
	Amount spent for books	: 6,63,042
	Amount spent for Journals	: 32,48,531
MCL :	Books & Journals purchase:	

The university taken to make the library a 'happening place' on campus:

Regular orientation programs are run by all these libraries time to time to all the students enrolled to the University about how to access and use print and electronic resources. Conducting of periodical information Literacy Programmes and Awareness Programmes like orientation and demo along with the Provision of A/c environment to readers for browsing to e- resources. Conducting regular training programmes, conferences and workshop to the library professional working in the affiliated colleges and other institutions libraries, Online awareness of the new arrivals so on.

The library to collect feedback form its users: The feedback analysed and used for the improvement of the library services:

MUL : Almost all the PG and Ph.D students of the University prefer to either access e-Resources in the OLIZ of the library or in their department. Periodical user studies are conducted and the results of the studies are published in the Conference Proceedings and Journals. According to the findings of these studies, the numbers of user nodes are improved considerably along with system speed and bandwidth. And according to the findings of the study more no. of reference and competitive examinations materials are acquired in recent years.

GCL : Conducting periodical user surveys and personal appraisals are the major criteria to get feed back to improve the library

TCL : Personal appraisals and continuous interactions with the users By providing all possible requirements of the Users

MCL : `Personal appraisals and continuous interactions with the users By providing all possible requirements of the Users Infrastructural development of the library:

Efforts			
MUL	**GCL**	**TCL**	**MCL**
• Renovation of entire library areas, • Establishment of Talking Library Section • International Conference on "Collection Development in Digital Environment (ICCDDE 2012) on 29-30 June 2012 • Digitisation of Theses collections in second phase work started • Upgradation of Book and membership database using SOUL Software 2.0 • Workshop on "INFLIBNET Regional Training Programme on Library Automation (IRTTPLA)" on 28 November to 2 December 2011. • Workshop on Library Network Protocols (LIBNETPRO 2011) On 25-26 March 2011	• Demonstrat ion of RFID tags in Guindy Campus Library • Electronic surveillanc e in Guindy Campus Library	• Intranet, • surveillance system, • digital library, • language laboratory, • virtual library	-

Chapter - IV

Analysis and Interpretation

4.1 GENERAL

The previous chapter provided the profile of libraries and particular in Madras University, Chennai, for the study. In this chapter analysis the data collected from the users in Madras University Library, The data was collected through the questionnaires (Appendix -1) and it has been analysed using statistical techniques and interpreted.

4.2. SAMPLE SIZE

The distribution of questionnaires among the users in Madras University Library has been analyzed based on the opinions and it is shown in the table.4.1.

Table 4.1 Distribution of Questionnaires among the Respondents

Sl. No	Questionnaires Distributed	%	Questionnaires Received	%
1	275	100	200	72.73

Source: Primary Data

Table 4.1 shows the distribution of questionnaires to the respondents from the Madras University Library in Tamil Nadu. Totally 275 questionnaires were distributed to the users in Madras University Library. Out of 275 200 replied after duly filled by the respondents. The response rate is 72.73 percent.

4.3 DEMOGRAPHIC DETAILS OF THE RESPONDENTS

The following table explains the demographic details such as gender, age, categories, domicile and faculty categories.

Table 4.2 Demographic Details of the Respondents

Sl. No	Description	Category	Questionnaire Distributed	%
1	Gender	Male	119	59.50
		Female	81	40.50
2	Respondents Category	Research Scholars	114	57.00
		Faculty	86	43.00
3	Age Category	Up to 30	38	19.00
		31-35	43	21.50
		36-40	43	21.50
		Above 40	76	38.00
4	Domicile	Urban	114	57.00
		Rural	86	43.00
6	Faculty Category	Science	86	43.00
		Social Science	76	38.00
		Humanities	38	19.00
		Total	**200**	**100.00**

Source: Primary Data

From the table 4.2 shows gender-wise distribution of respondents to use in Madras University Library. From the 200 respondents, 119 (59.50percent) are belongs to male and the remaining 81 (40.50per cent) are female.

Further the respondents were classified under their respondent category. Among the 200, 114 (57percent) of the respondents are Research scholars and the remaining 86 (43percent) of them were faculty members. It is concluded from the table that majority 114(57 percent) of the respondents were in the age category of Research Scholars.

The respondents were classified under their age frequency taken for the study. From the 200, 38 (19percent) were belong to the age group of up to 30", 43 (21.50percent) were in between the age group of „31-35",and „36-40" years of age and 76 (38percent) were at the age group of above 40 years. Majority of the respondents are above 40 years of age group.

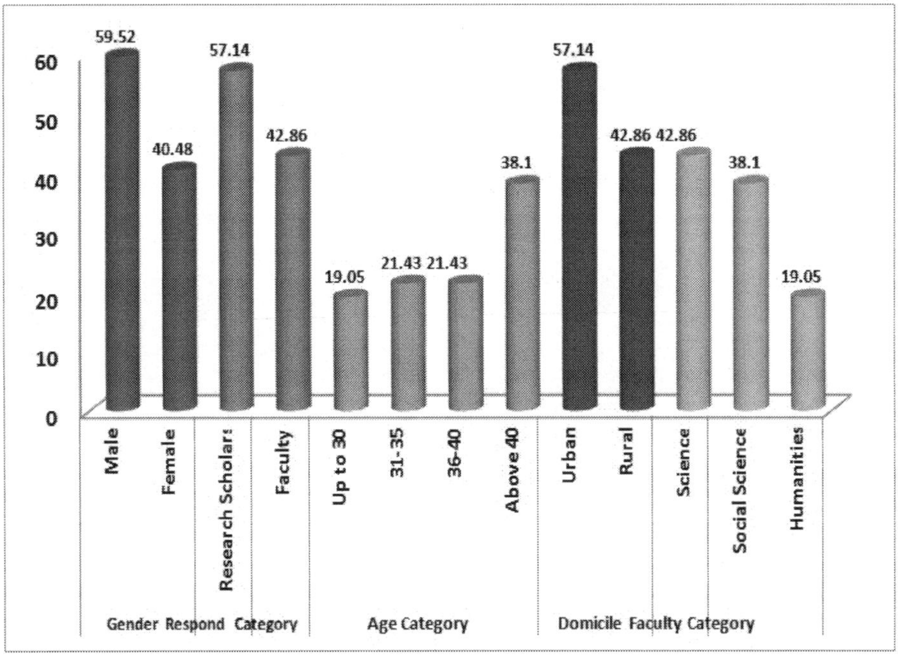

Figure 4.1 Demographic Details of the Respondents

Further the respondents were classified under their domicile. Among the 200, 114 (57percent) of the respondents were from urban and 86 (43percent) were from rural areas. It is studied from the table that majority of the respondents 57percent were in the category of urban.

Followed by, among the 200 respondents, 86 (43percent) respondents comes from from category of Science Department, 73 (38percent) were from Department of Social Science and the remaining 38 (19percent) respondents were from humanities. It is reveals that majority of the responses 43 percent from the Department of Science.

The sample library users of the respondents have been classified into two categories namely research scholars and faculty members for analytical purposes. This group is the yard stick to measure the information and literacy conditions of library users through Madras University Library.

4.4 FREQUENCY OF VISIT TO THE UNIVERSITY LIBRARY

The frequency of visit to the library among the users in the university library has been analyses based on the opinion of the respondents and it is shown in the table 4.3.

Table 4.3 Frequency of visit to the University library

Sl. No	Frequency	Respondent Category		Total Respondents
		Research Scholars	Faculty	
1	Daily	27 (32.37)	28 (32.59)	65 (32.47)
2	Twice a week	36 (31.50)	31 (36.09)	67 (33.47)
3	Weekly	29 (25.37)	23 (26.77)	52 (25.97)
4	Fortnightly	10 (8.75)	2 (2.33)	12 (12.53)
5	Monthly	2 (2.01)	2 (2.33)	4 (12.53)
Total		**114 (100.00)**	**86 (100.00)**	**200 (100.00)**

Source: Primary Data

Note : Figure in brackets is percentages to total.

It is revealed from the table 4.3, shows that the frequency of visit to the library among the users in the university library. The frequency was categories as follows:

„Daily", „ Twice a week", „ Weekly", „ Fortnightly" and „ Monthly". From the category of Research Scholars, 27 (32.37percent) of the respondents visit the library daily, 36 (31.50percent) are visiting twice a week, 29 (25.37percent) visits weekly, 10 (8.75percent) visits fortnightly and the remaining 2 (2.01percent) visits monthly. In the category of faculty members, 28 (32.59percent) of the respondents are visiting daily, 31 (36.09percent) of the respondents are visiting twice a week, 23 (26.77percent) of the respondents visits weekly, 2 (2.33percent) of the respondents visits fortnightly and monthly once. It is revealed from the table the majority of the respondents 67 (33.47percent) visit the library twice a week by the Research Scholars and Faculty members.

4.5 FREQUENCY OF TIME SPENT IN THE LIBRARY

The frequency of time spent in the library among the users in the university libraries has been analyses based on the opinion of the respondents and it is shown in the table 4.4.

Table 4.4 Frequency of Time Spent in the Library

Sl. No	Frequency	Respondent Category		Total Respondents
		Research Scholars	Faculty	
1	Up to 30 Minutes	7 (6.06)	24 (27.91)	31 (15.46)
2	30 Mts to 1hour	19 (16.69)	32 (37.21)	51 (25.52)
3	1 - 2 hours	48 (42.16)	23 (25.58)	70 (35.03)
4	2 - 3 hours	25 (21.96)	5 (5.81)	30 (15.01)
5	More than 3 hours	15 (13.13)	3 (3.49)	18 (8.98)
	Total	**114 (100.00)**	**86 (100.00)**	**200 (100.00)**

Source: Primary Data

Note : Figure in brackets is percentages to total.

It is inferred from the table 4.4, shows that the frequency of time spent in the library among the users in the Madras University library with their respondent category. From the category of Research Scholars, 48 (42.16%) of the respondents are spending 1 to 2 hours per day, 25 of the respondents (21.96%) spends 2 to 3 hours per day and 19 (16.69%) are spending 30 minutes to 1 hour. In the case of faculty members, 32 (37.21%) of the respondents spends 30 minutes to one hour, 24 (27.91%) of the respondents spends upto 30 minutes and 23 (25.58%) of the respondents are spending 1 to 2hours. It is identified from the table that majority 35.53 percent of the respondents are spending 1to 2 hours per day in the library.

4.6 THE PURPOSE OF VISITING THE LIBRARY - ALL

The study has been analyzed the purpose of visiting the library among the respondents. The five point scales of Never, Rarely, Sometimes, Often and More Often are used for this study. The mean, standard deviation and their rank for the purpose of visiting have been calculated and it is shown in the table 4.5.

Table 4.5 The Purpose of visiting the library

Sl. No	Description	Never	Rarely	Sometimes	Often	More often	Mean	SD	Ra nk
1	Read Newspapers and Read Periodicals	15	31	33	52	70	3.71	1.286	6
2	Prepare for Competitive Exams	15	2	51	59	73	3.87	1.162	5
3	General Knowledge and Employment News	2	6	4	94	93	4.28	0.915	2
4	Entertainment	6	12	37	61	83	4.05	1.014	3
5	Borrow Books	14	11	16	89	71	3.98	1.133	4
6	Academic Improvement and Research purpose	17	40	47	73	23	3.18	1.187	7
7	E-Resources and Others	6	5	20	47	123	4.36	0.879	1

Source: Primary Data

It is identified from Table 4.5 that the respondents are visiting the library more often for the purpose of „E-resources and others"(M=4.36). The respondents are visiting the library (M=4.28) for „general knowledge and employment news" and „entertainment" (4.05) are ranked as. Further, the respondents are visiting the library for the purpose of „academic improvement and research purpose". The mean value of all the variables range between 3.48 and 4.36. The standard deviation of opinion ranges between 0.879 and 1.286.

4.7 CATEGORY-WISE THE PURPOSE OF VISITING THE LIBRARY

The study has been further extended to category of the respondents for purpose of visiting the library among the users in University library. The mean, standard deviation and their rank for their purpose have been calculated and shown in table 4.6.

Table 4.6 Category-wise the purpose of Visiting the Library

Sl. No	Description	Research Scholars			Faculty			Chi-Square
		—	Std. Dev.	Rank	—	Std. Dev.	Rank	
1	Read Newspapers and Read Periodicals	3.58	1.114	6	3.89	1.201	5	28.546
2	Prepare for Competitive Exams	3.86	1.108	5	4.07	1.047	4	32.178
3	General Knowledge and Employment News	4.09	0.874	2	4.34	0.641	1	24.190
4	Entertainment	3.97	1.087	4	4.21	1.104	3	18.127
5	Borrow Books	4.01	1.208	3	3.79	0.988	6	7.495
6	Academic Improvement and Research purpose	3.24	1.163	7	3.31	1.118	7	18.574
7	E-Resources and Others	4.22	1.018	1	4.30	0.876	2	16.594

Source: Primary Data

Note : ¯ -Mean, SD-Standard Deviation, R-Rank.

As shown in the Table 4.6 that the respondents in the category of research scholars have given first priority for the purpose of „E-resources and others". „General Knowledge and Employment News" and „Borrow Books" are the second and third preference given by the respondents. The least preference has given for the need of „Academic Improvement and Research Purpose". The mean value of all the variables range between 3.24 and 4.22. The standard deviation of opinion range between 0.874 and 1.208.

Similarly the respondents in the category of faculty members has more often visited the purpose of „General Knowledge and Employment News". „E-resources and others" and „Entertainment" by the respondents. The respondents are often visited for „Academic Improvement and Research Purpose". The mean value of all the variables range between 4.34 and 3.31. The standard deviation of opinion range between 0.641 and 1.201.

Further, Chi square has been administered to identify the significance and the hypothesis "There is no significance difference between the respondent categories of research scholars and faculties towards the purpose of using library". The computed value has been shown in Table 4.6. The table value is 9.488 at 5% level of significance, the calculated value of all the variables are higher than the table value which indicated the variables are significant in their opinion between the respondent categories of research scholars and faculties towards the purpose of using libraries. Therefore the first hypothesis namely "There is no significance difference between the respondent categories of research scholars and faculties towards the purpose of using library" is insignificant and rejected.

4.8 OPINION TO DEVELOP THE GENERAL KNOWLEDGE AFTER GOING TO THE LIBRARY

The Opinion about the development of general knowledge after going to the library among the users in the university library has been analysed on the opinion of the respondents and it is shown in the table 4.7. The opinion was categories as follows: „Strongly Agree", „Agree", „No Opinion", „Disagree" and „Strongly Disagree".

Table 4.7 Opinion to Develop the General Knowledge after going to the Library

Sl. No	Description	Respondents Category		Total Respondents
		Research Scholars	Faculty	
1	Strongly Agree	30 (26.39)	23 (26.72)	53 (26.53)
2	Agree	34 (29.52)	32 (37.17)	66 (32.82)
3	No Opinion	10 (8.90)	5 (5.61)	15 (7.48)
4	Disagree	21 (18.47)	15 (17.42)	36 (18.02)
5	Strongly Disagree	19 (16.71)	11 (13.08)	30 (15.15)
	Total	**114 (100.00)**	**86 (100.00)**	**200 (100.00)**

Source: Primary Data

Note : Figure in brackets is percentages to total.

It is inferred from the table 4.7, shows that the opinion about the development of general knowledge after going to the library among the users in the University library. From the category of Research Scholars, 30 (26.39%) of the respondents are „strongly agreed", 34 (29.352percent) „agreed", "has No opinion" has towards the development of knowledge after going to the library. Followed by 10(8.90percent) has "No opinion" 21(18.47percent) are disagreed and 19(16.71percent) of the respondents are strongly disagreed regarding the development of Knowledge after going to the library. In case of faculty members 32(37.17percent) are agreed, 23(26.72percent) are strongly agreed towards the development of General Knowledge after visiting the library. Further, 5(5.61percent) has no opinion, 15(17.42percent) are disagree and 11 (13.08percent) has disagreed to development of General Knowledge after visiting the library. It is identified from the table that majority 32.82 per cent of the respondents „agreed" to the development of general knowledge after going to the Library.

4.9 PLACE OF ACCESSING THE ELECTRONIC INFORMATION

The place of accessing the electronic information among the users in the university library has been analysed based on the opinion of the respondents and it is shown in the table 4.8.

Table 4.8 Place of accessing the electronic information

Sl. No	Description	Respondents Category		Total Respondents
		Research Scholars	Faculty	
1	Central Library	42 (36.74)	37 (43.49)	79 (39.64)
2	Departmental Library	21 (18.37)	11 (12.77)	32 (15.96)
3	Hostel	15 (13.12)	21 (24.01)	36 (17.80)
4	Home	17 (14.68)	8 (9.28)	25 (12.36)
5	Cybercafe	20 (17.09)	9 (10.45)	29 (14.24)
	Total	**114 (100.00)**	**86 (100.00)**	**200 (100.00)**

Source: Primary Data

Note : Figure in brackets is percentages to total.

It is revealed from the table 4.8, shows that 42(36.74percent) of the Research Scholars are accessing the electronic information from the central library, 21 (18.37percent) of the respondents are accessing from the department library and 20 (17.09percent) are accessing from Cybercafe. In the case of faculty members, 37 (43.49percent) are accessing from Central Library, 21 (24.01percent) of the respondents are accessing from hostel and 11 (12.77percent) are accessing from Department Library. It is identified from the table that majority 39.64 per cent of the respondents are accessing the electronic information from Madras University Central Library..

4.10 LEVEL OF USING THE ELECTRONIC INFORMATION RESOURCES

The level of using the electronic information resources among the users in the university library has been analysed based on the opinion of the respondents and it is shown in the table 4.9.

Table 4.9 Level of using the Electronic Information Resources

Sl. No	Description	Respondents Category		Total Respondents
		Research Scholars	Faculty	
1	Beginner	10 (9.05)	5 (5.37)	15 (7.47)
2	Average	65 (56.85)	41 (47.90)	106 (53.02)
3	Expert	39 (31.11)	40 (46.73)	79 (39.51)
	Total	**114 (100.00)**	**86 (100.00)**	**200 (100.00)**

Source: Primary Data

Note : Figure in brackets is percentages to total.

It is revealed from the table 4.9, that the level of using the electronic information resources among the users in the university library analysed with their respondents category. From the category of Research Scholars, 10 (9.05percent) are in the level of beginner, 65 (56.85percent) are in the level of average and the remaining 39 (31.11percent) of the respondents were in the level of expert. In the case of faculty member, 5 (5.37percent) are in the level of beginner and 41 (47.90percent) are in the level of average and the remaining 40 (46.73percent) of the respondents are in the level of expert. It is identified from the table the majority of the respondents 106 (53.02percent) of the respondents are average level of using the electronic information resources.

4.11 FREQUENCY OF USING THE ELECTRONIC INFORMATION RESOURCES

The frequency of using the electronic information resources among the users in the university library has been analyses based on the opinion of the respondents and it is shown in the table 4.10.

Table 4.10 Frequency of using the Electronic Information Resources

Sl. No.	Description	Respondents Category		Total Respondents
		Research Scholars	Faculty	
1	Rarely	3 (2.63)	3 (3.99)	6 (3.22)
2	Frequently	61 (53.51)	41 (47.43)	102 (50.89)
3	More frequently	50 (43.86)	42 (48.58)	92 (45.90)
	Total	**114 (100.00)**	**86 (100.00)**	**200 (100.00)**

Source: Primary Data

Note : Figure in brackets is percentages to total.

It is revealed from the table 4.10, shows that the frequency of using the Electronic Information Resources. From the category of Research Scholars, 3 (2.63%percent are used rarely, 61 (43.86percent) are used frequently and 50 (43.86percent) are used more frequently. In the case of faculty member 3 (3.99percent) are used rarely and 42 (48.58percent) are used more frequently. It is identified from the table majority of the respondents 102 (50.89percent) were in the category of research scholars and they are frequently used the electronic information resources.

4.12 TIME SPEND FOR ELECTRONIC INFORMATION PER DAY

The time spend for electronic information per day among the users in the university library has been analysed based on the opinion of the respondents and it is shown in the table 4.11.

Table 4.11 Time spend for Electronic Information Per day

Sl. No	Description	Respondents Category		Total Respondents
		Research Scholars	Faculty	
1	Below 30 minutes	5 (4.39)	2 (2.33)	7 (3.51)
2	30 Mts to 1hour	4 (3.43)	4 (4.29)	8 (3.80)
3	1 - 2 hours	20 (17.56)	13 (15.17)	33 (16.53)
4	2 - 3 hours	48 (42.14)	14 (16.34)	62 (31.05)
5	More than 3 hours	37 (32.48)	53 (61.86)	90 (45.09)
	Total	**114 (100.00)**	**86 (100.00)**	**200 (100.00)**

Source: Primary Data

Note : Figure in brackets is percentages to total.

The table 4.11 shows that, the time spent for electronic information per day among the users in the university library of the respondents. From the category of Research Scholars, 48 (42.14percent) respondents are spending in 2-3 hours for Electronic Information followed by 37 (32.48percent) of the respondents are spending more than 3 hours and 20 (17.56percent) of the respondents are spending 1-2 hours for spending Electronic Information. In the case of Faculty Members, 53 (61.86percent) respondents are spending more than 3 hours. Followed by 14 (16.34percent) of the respondents are spending 2-3 hours and 13 (15.17percent) of the respondents are spending 1-2 hours. It is identified from the above table that majority 45.09 percent of the respondents are spending more than 3 hours per day for seeking the electronic information.

4.13 METHODS TO LEARN THE USE OF E- RESOURCES

The methods to learn the use of E- resources among the users in the university library has been analysed based on the opinion of the respondents and it is shown in the table 4.12.

Table 4.12 Methods to learn the use of E- resources

Sl. No	Description	Respondents Category		Total Respondents
		Research Scholars	Faculty	
1	Guidance from friends	14 (12.28)	15 (17.44)	29 (14.50)

Sl. No	Description	Respondents Category		Total Respondents
		Research Scholars	Faculty	
2	Trial & Error Method/ Self Instruction	22 (19.30)	18 (20.93)	40 (20.00)
3	Formal training	24 (21.05)	8 (9.30)	32 (16.00)
4	By Attending Courses	27 (23.68)	11 (12.79)	38 (19.00)
5	From teachers	24 (21.05)	31 (36.05)	55 (27.50)
6	From Librarians	3 (3.63)	3 (3.49)	6 (3.00)
	Total	**114 (100.00)**	**86 (100.00)**	**200 (100.00)**

Source: Primary Data

Note : Figure in brackets is percentages to total.

It can be revealed from the table 4.12 that the methods to learn the use of E- resources among the users in the university library. Among the Research Scholars, 27 (23.68percent) of the respondents are learning by attending courses, 24 (21.05percent) learns by attending formal training and from teachers. In the case of Faculty Members 31 (36.05percent) of the respondents are learning the use of e-resources from teachers, and 18 (19.30percent) learns through trail & error method and self instruction. It is identified from the table that the majority (27.50percent) of the respondents learns through teachers regarding the the use of E-resources.

4.14 TYPES OF E-RESOURCES ACCESSED IN LIBRARIES

The study has been analysed the types of E-resources accessed by the users in University library. The five point scales of never, rarely, sometimes, often and more often were used for the study. The Mean, Standard Deviation and their Rank for the types of E-resources accessed have been calculated and it is shown in the table 4.13.

Table 4.13 Types of E-resources accessed in Libraries

Sl. No	Description	Never	Rarely	Some-times	Often	More often	–	Std. Dev.	Rank
1	Search engines	2	9	16	79	93	4.54	0.847	2
2	Websites	6	9	31	49	105	4.32	1.023	3

Sl. No	Description	Never	Rarely	Some-times	Often	More often	–	Std. Dev.	Rank
3	E-mail	6	7	14	23	151	4.68	0.988	1
4	E- journals/e-Books	8	9	22	74	87	4.21	1.024	4
5	Databases	11	8	36	75	70	3.90	1.066	6
6	Professional societies/asso-ciation	7	14	17	74	88	4.14	1.104	5
7	List serve/discus-sion group	6	22	40	82	51	3.79	1.101	7
8	Bulletin board services	9	23	59	63	46	3.63	1.097	8

Source: Primary Data

It is inferred from Table 4.13 that the respondents are more often accessing E- resources through „E-mail". „Search engines" and „Websites". The respondents are sometimes accessing resources through of „Bulletin board services". The mean value of all the variables range between 3.63 and 4.68. It is inferred that all the eight variables lie between „More often to Rarely". The standard deviation of opinion ranges between 0.847 and 1.104.

4.15 TYPES OF E-RESOURCES ACCESSED IN LIBRARIES

The study has been further extended to respondent category for the types of E-resources accessed by the users in university library. The Mean, Standard Deviation and their Rank for the types of E-resources accessed have been calculated and it is shown in the table 4.14.

Table 4.14 Types of E-resources accessed in Libraries

Sl. No	Description	Research Scholars			Faculty			Chi-Square
		—	Std. Dev.	Rank	—	Std. Dev.	Rank	
1	Search engines	4.34	0.881	2	4.30	0.914	4	14.324
2	Websites	4.06	1.108	5	4.57	0.874	2	24.567

Sl. No	Description	Research Scholars			Faculty			Chi-Square
		—	Std. Dev.	Rank	—	Std. Dev.	Rank	
3	E-mail	4.61	1.024	1	4.59	0.948	1	23.257
4	E- journals/e-Books	4.18	1.103	4	4.39	1.001	3	15.624
5	Databases	3.97	1.147	6	3.94	1.014	7	19.023
6	Professional societies/ association	4.23	1.155	3	4.23	0.947	5	26.359
7	List serve/discussion group	3.87	1.087	7	4.04	1.002	6	38.457
8	Bulletin board services	3.74	1.113	8	3.71	1.054	8	32.156

Source: Primary Data

Note $\bar{}$ -Mean, SD-Standard Deviation, R-Rank

It is revealed from Table 4.14 that the respondents in the category of Research Scholars are more often accessing E-resources through „E-mail". „Search engines" and „Professional societies/association" The respondents are sometimes accessing E- resources through „Bulletin board services". The mean value of all the variables range between 3.74 and 4.61. It is inferred that all the eight variables lie between „More often to rarely". The standard deviation of opinion range between 0.881 and 1.155.

The respondents in the category of faculty members are more often accessing the e-resource through „E-mail", „Websites" and „E- journals/e-Books". The respondents are sometimes accessing the e-resource through „Bulletin board services". The mean value of all the variables range between 3.71 and 4.59. It is inferred that all the eight variables lie between „More often to Rarely". The standard deviation of opinion ranges between 0.874 and 1.054.

Further, Chi square has been administered to identify the significance. The calculated value has been shown in Table 4.14. The table value is 9.488 at 5percent level of significance, the computed value is higher than the table value which indicated the variables are significant in their opinion between the respondent categories towards the types of E-resources accessed by the users.

4.16 RATING THE PURPOSE OF USING THE ELECTRONIC INFORMATION RESOURCES

The study has been analysed the purpose of using the electronic information resources by the users in university library. The five point scales of not at all, when required, once in a week, twice in a week, daily were used for the study. The Mean, Standard Deviation and their Rank for the types of E-resources accessed have been calculated and it is shown in the table 4.15.

Table 4.15 Rating the Purpose of using the Electronic information resources

Sl. No	Description	Not at all	When required	Once in a week	Twice in a week	Daily	—	SD	Rank
1	To keep abreast with the latest development	2	7	33	41	118	4.48	0.984	1
2	For study/ research	2	25	11	66	96	4.14	1.102	6
3	For career development	1	4	22	97	76	4.33	0.874	3
4	For teaching preparation	2	4	21	86	87	4.42	0.924	2
5	For writing books/ research articles	4	3	36	71	87	4.21	1.081	5
6	For communi-cation (e-mail)	3	14	20	81	82	4.10	1.024	7
7	Professional Development	2	25	38	77	58	3.95	1.114	8
8	E-shopping	5	20	53	54	68	3.78	1.078	9
9	Entertainment	5	5	33	59	98	4.28	1.101	4

Source: Primary Data

It is identified from table 4.15 the respondents are using the electronic information Daily in order to keep the abreast with the latest development", „For teaching preparation" and „For career development". The respondents are using the electronic information twice in a week for E-shopping". The mean value of all the variables range between 3.78 and 4.48. It can be inferred that all the nine variables lie between „twice in a week" and „daily". The standard deviation of opinion ranges between 0.874 and 1.114.

4.17 PURPOSE OF USING THE ELECTRONIC INFORMATION RESOURCES

The study has been further extended to respondents for the purpose of using the electronic information resources by the users in university library. The Mean, Standard Deviation and their Rank for the purpose of using the electronic information resources have been calculated and it is shown in the table 4.16.

Table 4.16 Purpose of using the Electronic Information Resources

Sl. No	Description	Research Scholars			Faculty			Chi-Square
		—	Std. Dev.	Rank	—	Std. Dev.	Rank	
1	To keep abreast with the latest development	4.44	0.978	1	4.48	0.857	2	5.127
2	For study/ research	3.98	1.103	7	4.59	0.767	1	19.456
3	For career development	4.36	0.841	2	4.19	1.002	6	31.287
4	For teaching preparation	4.23	0.947	4	4.39	0.817	3	4.156
5	For writing books/ research articles	4.28	0.912	3	4.09	0.899	7	13.298
6	For communication (e- mail)	4.10	1.016	6	4.26	0.788	5	24.358
7	Professional Development	3.74	1.047	9	4.04	0.897	8	32.687
8	E-shopping	3.89	1.122	8	3.68	1.036	9	16.459
9	Entertainment	4.17	1.068	5	4.31	1.102	4	14.236

Source: Primary Data

Note : ̄-Mean, SD-Standard Deviation

It is inferred that the respondents in the category of Research Scholars are using the E-resources daily in order „To keep abreast with the latest development", „For career development" and „For writing books/ research articles". The respondents are using E-resources once in a week for „Professional Development". The mean value of all the variables range between 3.74 and 4.44. It can be inferred that all the nine variables lie between „Daily to Not at all". The standard deviation of opinion ranges between 0.841 and 1.122.

The respondents in the category of Faculty Members are using Daily for the purpose of „For Study/ Research", „To keep abreast with the latest development" and „For teaching preparation". The respondents are using once in a week for „E- shopping". The mean value of all the variables range between 3.68 and 4.59. It can be inferred that all the nine variables lie between „Daily to Not at all". The deviation of opinion ranges between 0.788 and 1.102

Further, Chi square has been administered to identify the significance. The calculated value has been shown in Table 4.16. The table value is 9.488 at 5percent level of significance, the computed value of most of the variables were higher than the table value which indicated the variables are significant in their opinion between the respondent categories towards the purpose of using the electronic information resources by the users except the variables of „To keep abreast with the latest development" (x^2 =5.127) and „For teaching preparation (x^2 =4.156)

4.18 RATING THE FEATURES OF ELECTRONIC INFORMATION

The study has been analysed the rating the features of electronic information by the users in university library. The five point scales of Poor, No comments, Satisfaction, Good, Very Good are used for the study. The Mean, Standard Deviation and their Rank for the features of Electronic information have been calculated and it is shown in the table 4.17.

Table 4.17 Rating the features of Electronic Information

Sl. No	Description	Very Poor	Poor	No Comments	Good	Very Good	—	Std. Dev.	Rank
1	Uniqueness	6	11	9	54	119	4.39	1.122	4
2	Accessibility	6	13	10	53	118	4.27	1.103	5
3	Accuracy	6	13	12	54	115	4.10	1.068	7
4	Availability	6	13	6	36	139	4.53	1.034	2
5	Reliability	6	13	12	64	104	4.02	1.046	9
6	Flexibility	6	13	11	45	125	4.46	1.101	3
7	Stability	6	13	12	62	106	4.06	1.040	8

Sl. No	Description	Very Poor	Poor	No Comments	Good	Very Good	—	Std. Dev.	Rank
8	Timeliness	1	16	9	71	105	4.11	0.987	6
9	Ease of use	1	11	6	56	127	4.68	0.903	1

Source: Primary Data

It is evident from Table 4.17 that the respondents feels that the feature of „Ease of use", „Availability" and „Flexibility" are very good in Electronic Information. The respondents feels that Electronic Information also very good in terms of „Reliability". The mean value of all the variables range between 4.02 and 4.68. It is inferred that all the nine variables lie on „Very Good". The standard deviation of opinion ranges between 0.903 and 1.122.

4.19 RATING THE FEATURES OF ELECTRONIC INFORMATION

The study has been further extended to the respondents for the rating the features of electronic information by the users in university library. The Mean, Standard Deviation and their Rank for the features of the electronic information resources have been calculated and it is shown in the table 4.18.

Table 4.18 Rating the features of Electronic Information

Sl. No	Description	Research Scholars			Faculty			Chi- Square
		—	Std. Dev.	R	—	Std. Dev.	R	
1	Uniqueness	4.28	1.089	4	4.49	0.948	5	12.358
2	Accessibility	4.19	1.044	7	4.55	1.016	4	20.489
3	Accuracy	4.17	1.110	8	4.38	0.948	6	14.587
4	Availability	4.63	1.053	1	4.76	1.031	2	20.365
5	Reliability	4.21	1.114	6	4.14	0.956	9	40.583
6	Flexibility	4.24	1.088	5	4.67	0.987	3	17.487
7	Stability	4.12	1.108	9	4.29	0.956	8	16.580

Sl. No	Description	Research Scholars			Faculty			Chi- Square
		—	Std. Dev.	R	—	Std. Dev.	R	
8	Timeliness	4.38	0.948	3	4.31	0.986	7	13.402
9	Ease of use	4.51	0.954	2	4.81	0.758	1	23.607

Source: Primary Data

Note : ¯-Mean, SD-Standard Deviation, R-Rank

It can be analysed from Table 4.18 that the respondents in the category of „Research Scholar" are very good regarding the features of „Availability" „Ease of use" and „Timeliness". It is also very good to access Electronic Information in terms of Stability". The mean value of all the variables ranges between 4.63 and 4.12. It is inferred that all the nine variables lies on „Very Good". The standard deviation of opinion ranges between 0.954 and 1.114.

The respondents in the category of „Faculty Members" views that Electronic Information features are very good in terms of „Ease of use", „Availability" and Flexibility" Further, it is very good to access Electronic Information as it is "Reliable". The mean value of all the variables ranges between 4.81 and 4.14. It can be inferred that all the nine variables lies on „Very Good". The standard deviation of opinion ranges between 0.758 and 1.031.

Further, Chi square has been administered to identify the significance. The calculated value has been shown in Table 4.18. The table value is 9.488 at 5percent level of significance, the calculated value of all the variables were higher than the table value which indicated the variables are significant in their opinion between the respondent categories towards the features of electronic information by the users.

4.20 THE ACCEPTANCE OF ACCESSING INFORMATION IN A DIGITAL LIBRARY

The acceptance of accessing information in a digital library among the users in the university libraries has been analyses based on the opinion of the respondents and it is shown in the table 4.19.

Table 4.19 The Acceptance of Accessing Information in a Digital Library

Sl. No	Description	Respondents Category		Total Respondents
		Research Scholars	Faculty	
1	Saves more physical space	18 (15.79)	4 (4.65)	22 (11.00)
2	Library services shift from 'just in case' to 'just in time'	3 (2.63)	2 (2.33)	5 (2.50)
3	Access from anywhere at any time	21 (18.42)	22 (25.58)	43 (21.50)
4	Availability of information to all users simultaneously	15 (13.16)	24 (27.91)	39 (19.50)
5	Easy retrieval	29 (25.44)	28 (32.56)	57 (28.50)
6	No Maintenance	22 (19.30)	3 (3.49)	25 (12.50)
7	Enables networking of various internal resources	4 (3.51)	2 (2.33)	6 (3.00)
8	Networking of other libraries	2 (1.75)	1 (1.16)	3 (1.50)
	Total	**114 (100.00)**	**86 (100.00)**	**200 (100.00)**

Source: Primary Data

Note : Figure in brackets is percentages to total.

It is revealed from the table 4.19, shows that the acceptance of accessing information in a digital library among the users in the university library with their respondents. From the category of Research Scholars, 18 (15.79percent) accepts that „saves more physical space", 3 (2.63percent) accepts that „library services shift from 'just in case' to 'just in time', 15 (13.16percent) accepts to use Electronic Information for „availability of information to all users simultaneously" and 28(25.44percent) are using for „easy retrieval". In the case of „Faculty Members", 2 (2.33percent) are using the Library services shift from 'just in case' to 'just in time', 22 (25.58percent) accepts to access from anywhere at any time" and 28 (32.56percent) accepts to access for „easy retrieval". It is identified from the table that majority of the respondents 57 (28.50percent) were in the total respondents and they are accepting to access for „easy retrieval" of accessing information in a digital library.

4.21 THE EXTENT OF SATISFACTION WITH ELECTRONIC INFORMATION RESOURCES

The study has been analysed the extent of satisfaction with electronic information resources by the users in universities library. The five point scales of Dissatisfied, No Comments, Somewhat satisfied, Satisfied, Highly Satisfied are used for the study. The Mean, Standard Deviation and their Rank for the extent of satisfaction with electronic information resources have been calculated and it is shown in the table 4.20.

Table 4.20 The extent of Satisfaction with Electronic Information Resources

Sl. No	Factors	Dissatisfied	No Comments	Somewhat Satisfied	Satisfied	Highly Satisfied	-	Std. Dev.	Rank
1	Internet Sources	2	3	20	59	116	4.58	0.987	1
2	Subscribed Databases	11	37	10	64	78	3.81	1.245	6
3	E- Books	3	16	59	44	79	3.88	1.088	5
4	A/V Materials	1	36	25	53	84	3.91	1.175	4
5	E- Journals	1	18	9	96	77	4.39	0.906	2
6	Ask a Librarians	34	30	33	53	50	3.74	1.318	7
7	CD-ROM Databases	39	42	14	63	43	3.69	1.384	9
8	E-Catalogue	30	32	39	60	39	3.54	1.299	8
9	Open source Databases	1	23	35	64	77	4.16	1.057	3

Source: Primary Data

It can be identified from Table 4.20 that the respondents are highly satisfied towards for factors like „Internet Sources", „E- Journals" and „Open source Databases" The respondents are somewhat satisfied with Electronic Information Resources like CD-ROM Database. The mean value of all the variables ranges between 3.54 and 4.58. It can be inferred that all the nine variables lies between

„Highly Satisfied" to „ Highly Dissatisfied". The standard deviation of opinion ranges between 0.906 and 1.384.

4.22 THE EXTENT OF SATISFACTION WITH ELECTRONIC INFORMATION RESOURCES

The study has been further extended to respondent category for the extent of satisfaction with electronic information resources by the users in University library. The Mean, Standard Deviation and their Rank for the extent of satisfaction with Electronic information resources have been calculated and it is shown in the table 4.21.

Table 4.21 The extent of Satisfaction with Electronic Information Resources

Sl. No	Description	Research scholar			Faculty			Chi-Square
		–	Std. Dev.	R	–	Std. Dev.	R	
1	Internet Sources	4.63	0.798	1	4.42	0.899	2	12.598
2	Subscribed Databases	3.82	1.206	6	4.09	1.236	4	24.658
3	E- Books	3.98	1.152	4	4.01	1.078	5	13.021
4	A/V Materials	4.08	1.123	3	3.97	1.224	6	12.785
5	E- Journals	4.19	0.968	2	4.59	0.817	1	26.357
6	Ask a Librarians	3.76	1.421	7	3.36	1.278	9	23.158
7	CD-ROM Databases	3.34	1.498	9	3.83	1.216	7	16.957
8	E-Catalogue	3.66	1.218	8	3.69	1.277	8	11.784
9	Open source Databases	3.89	1.231	5	4.37	0.728	3	13.598

Source: Primary Data

Note : $\bar{}$-Mean, SD-Standard Deviation, R-Rank

It is revealed from Table 4.21 that the respondents in the category of „Research Scholars" Highly Satisfied with e-resource of „Internet Sources", „E- Journals" and „A/V Materials". The respondents are also somewhat satisfied with e- resources of „CD-ROM Databases". The mean value of all the variables ranges between 3.36 and 4.63. It can be inferred that all the nine variables lies between „Highly Satisfied" and „ Highly Dissatisfied". The standard deviation of opinion ranges between 0.798 and 1.498.

The respondents in the category of „Faculty Members" are highly satisfied with the e-resource of „E- Journals", „Internet Sources" and „Open source Databases" The respondents are somewhat satisfied with „Ask a Librarians". The mean value of all the variables range between 3.36 and 4.59. It is inferred that all the nine variables lie between „Some what satisfied" and „ Highly Satisfied". The standard deviation of opinion ranges between 0.817 and 1.278.

Further, Chi square has been administered to identify the significance and the hypothesis "There is no significance difference between the respondent categories of research scholars and faculties towards the extent of satisfaction with electronic information resources". The calculated value has been shown in Table 4.21. The table value is 9.488 at 5% level of significance, the calculated value of all the variables are higher than the table value which indicated the variables are significant in their opinion between the respondents categories towards the extent of satisfaction with electronic information resources by the users. Therefore the second hypothesis namely "There is no significance difference between the respondent categories of research scholars and faculties towards the extent of satisfaction with electronic information resources" is insignificant and rejected.

4.23 ACCESSIBILITY OF UGC- ESS E-JOURNALS IN LIBRARY

The accessibility of UGC- eSS E-Journals in library among the users in the university library has been analysed based on the opinion of the respondents and it is shown in the table 4.22.

It can be revealed from the table 4.22, shows that the Accessibility of UGC- eSS E-Journals in Library among the users in the university library with their category of the Gender.

Table 4.22 Accessibility of UGC- eSS E-Journals in Library

Sl. No	Description	Respondents Category		Total Respondents
		Research Scholars	Faculty	
1	Yes	98 (85.96)	74 (86.05)	172 (86.00)
2	No	16 (14.04)	12 (13.95)	28 (14.00)
	Total	114 (100.00)	86 (100.00)	200 (100.00)

Source: Primary Data

Note : Figure in brackets is percentages to total.

It is revealed from the table 4.22, shows that the accessibility of UGC- eSS E-Journals among the users in the university library with respondents. It found that 98(85.96) per cent of the respondents are more awareness and also the accessed the UGC eSS E-journals in Madra University library.

4.24 PURPOSE OF USING UGC- ESS E- JOURNALS

The purpose of using UGC- eSS E- Journals among the users in the university library has been analysed based on the opinion of the respondents and it is shown in the table 4.23.

Table 4.23 Purpose of using UGC- eSS E- Journals

Sl. No	Description	Respondents Category		Total Respondents
		Research Scholars	Faculty	
1	Research Work	22 (19.30)	24 (27.91)	46 (23.00)
2	Publishing articles/ books	17 (14.91)	23 (26.74)	40 (20.00)
3	Keeping up-to-date information	26 (22.81)	19 (22.09)	45 (22.50)
4	Finding relevant information in the area of specialization	17 (14.91)	11 (12.79)	29 (14.00)
5	UGC NET	32 (28.07)	9 (10.47)	41 (20.50)
	Total	114 (100.00)	86 (100.00)	200 (100.00)

Source: Primary Data

Note : Figure in brackets is percentages to total.

Table 4.23 shows that the purpose of using UGC- eSS E- Journals among the users in the university library with their respondent category. From the category of „research scholars" 22 (19.30percent) are using „Research Work", 17 (14.91percent) are „Publishing articles/ books', 26 (22.81percent) are „Keeping up-to-date information" 32(28.07) are using UGC NET and 17 (14.91percent) are for „Finding relevant information in the area of specialization". In the case of „faculty", 23 (26.74percent) are for „Publishing articles/ books" and 19 (22.09percent) are for „Keeping up-to-date information". It is identified from the table the majority 23 per cent of the respondents are using the publishing articles / books.

4.25 THE PROBLEMS FACED IN USING UGC- ESS E- JOURNALS

The study has been analysed the Problems faced in using UGC- eSS E- Journals by the users in university library. The five point scales of Strongly Disagree, Disagree, No Opinion, Agree and Strongly Agree are used for the study. The Mean, Standard Deviation and their Rank for the problems have been calculated and it is shown in the table 4.24.

Table 4.24 The Problems faced in using UGC- eSS E- Journals

Sl. No	Description	SDA	DA	NO	A	SA	—	Std. Dev.	R
1	Slow Internet Bandwidth	2	1	13	21	163	4.79	0.784	1
2	Lack of sufficient nodes in university	1	3	23	29	143	4.62	0.845	2
3	Frequent Power cut	1	1	15	124	59	4.21	0.769	7
4	Restricted access	1	1	26	57	114	4.44	0.864	4
5	Technical Problems (Server Down)	1	20	32	63	85	4.09	1.032	8
6	Content searching is difficult as compared to search Engine	3	2	24	45	126	4.58	0.914	3
7	Lack of assistance by library staff	2	2	14	111	71	4.38	0.718	5
8	Lack of knowledge to use	2	2	27	90	79	4.27	0.732	6

Source: Primary Data

Note: SDA- Strongly Disagree, DA- Disagree, NO- No Opinion, A- Agree, SA- Strongly Agree.

It is revealed from Table 4.24 that the respondents strongly agree for the problem of „Slow Internet Bandwidth". „Lack of sufficient nodes in university" and„ Content searching is difficult as compared to search Engine". The respondents also strongly agree with the „Technical Problems (Server Down)". The mean value of all the variables range between 4.09 and 4.79. It can be inferred that all the eight variables lie on „Strongly Agree". The standard deviation of opinion ranges between 0.718 and 1.032.

4.26 THE PROBLEMS FACED IN USING UGC- ESS E- JOURNALS

The study has been further extended to respondent category for the problems faced in using UGC- eSS E- Journals by the users in universities library. The Mean, Standard Deviation and their Rank for the problems faced in using UGC- eSS E- Journals have been calculated and it is shown in the table 4.25.

Table 4.25 The problems faced in using UGC- eSS E- Journals

Sl. No	Description	Research Scholars			Faculty			Chi-Square
		—	Std. Dev.	Rank	—	Std. Dev.	Rank	
1	Slow Internet Band-width	4.80	0.784	1	4.78	0.681	1	16.157
2	Lack of sufficient nodes in university	4.77	0.935	2	4.67	0.841	2	16.257
3	Frequent Power cut	4.47	0.647	5	4.19	0.794	8	26.182
4	Restricted access	4.63	0.798	3	4.29	0.867	4	17.367
5	Technical Problems (Server Down)	3.87	1.034	8	4.39	0.915	5	24.781
6	Content searching is difficult as compared to search Engine	4.57	0.879	4	4.55	0.783	3	18.469
7	Lack of assistance by library staff	4.42	0.729	6	4.28	0.728	6	19.014
8	Lack of knowledge to use	4.33	0.768	7	4.22	0.733	7	18.459

Source: Primary Data

Note : ¯-**Mean, SD-Standard Deviation, R-Rank, Degrees of Freedom: 4, Table Value – 9.488.**

It is revealed from Table 4.25 that the respondent category of „research scholars" strongly agree with the problem„ Slow Internet Bandwidth", „ Lack of sufficient nodes in university Restricted access" and „Technical Problems (Server Down)". The mean value of all the variables ranges between

3.87 and 4.80. It can be inferred that all the eight variables lies on „Strongly Agree". The standard deviation of opinion ranges between 0.647 and 1.034.

The respondent in the category of „faculty" are strongly agree with problem of „Slow Internet Bandwidth", „Lack of sufficient nodes in university", „Content searching is difficult as compared to search Engine", and „Frequent Power cut". The mean value of all the variables range between 4.19 and 4.78. It can be inferred that all the eight variables lie on „Strongly Agree". The standard deviation of opinion ranges between 0.681and 0.915.

Further, Chi square has been administered to identify the significance. The calculated value has been shown in Table 4.25. The table value is 15.507 at 5percent level of significance, the computed value of all the variables are higher than the table value which indicated the variables are significant in their opinion between the faculty categories towards the problems faced in using UGC- eSS E- Journals by the users in university library.

4.27 OPINION ON UGC-ESS JOURNALS IN DEVELOPING THE ACADEMIC SKILLS

The Opinion on UGC-eSS journals in developing the academic skills among the users in the university libraries has been analyses based on the opinion of the respondents and it is shown in the Table 4.26.

Table 4.26 Opinion on UGC-eSS Journals in Developing the Academic Skills

Sl. No.	Description	Respondents Category		Total Respondents
		Research Scholars	Faculty	
1	Strongly Disagree	1 (0.88)	3 (3.49)	4 (2.00)
2	Disagree	4 (3.51)	5 (5.81)	9 (4.50)
3	No Opinion	13 (11.40)	10 (11.63)	23 (11.50)
4	Agree	65 (57.02)	40 (46.51)	105 (52.50)
5	Strongly Agree	31 (27.19)	28 (32.56)	59 (29.50)
	Total	114 (100.00)	86 (100.00)	200 (100.00)

Source: Primary Data

Note : Figure in brackets is percentages to total.

The table 4.26, shows that the opinion about the on UGC-eSS journals in developing the academic skills among the users in the university libraries with their respondent category. From the category of „Research scholars", 31 (27.19 percent) are„ strongly agreed", 65 (57.02percent) are „Agreed" and 13 (11.40%) are „No Opinion" towards developing the Academic skills. In the case of „Faculty Members", 28 (32.56percent) are „Strongly Agreed", 40 (46.51%) are „Agreed" and 5 (5.81percent) are „Disagreed" for developing the Academic skills. It is identified from the table the majority of the respondents 105 (52.50percent) are „Agreeing that the UGC-eSS journals helps in developing the academic skills.

4.28 THE LEVEL OF SATISFACTION IN USING UGC-ESS CONSORTIUM

The level of satisfaction in using UGC-eSS consortium among the users in the university library has been analysed on the respondents and it is shown in the table 4.27.

Table 4.27 The level of Satisfaction in using UGC-eSS Consortium

Sl. No	Description	Respondents Category		Total Respondents
		Research Scholars	Faculty	
1	Very satisfied	26 (22.81)	11 (12.79)	37 (18.50)
2	Satisfied	59 (51.75)	30 (34.88)	89 (44.50)
3	No Opinion	15 (14.00)	27 (31.40)	43 (21.50)
4	Dissatisfied	6 (5.26)	15 (17.44)	21 (10.50)
5	Very Dissatisfied	7 (6.14)	3 (3.49)	10 (5.00)
	Total	**114 (100.00)**	**86 (100.00)**	**200 (100.00)**

Source: Primary Data

Note : Figure in brackets is percentages to total.

It is inferred from the table 4.27, shows that the level of satisfaction in using UGC-eSS consortium among the users in the university libraries with their respondent category. From the category of „Research scholars", 26 (22.81percent) are „Very Satisfied", 59 (51.75percent) are „Satisfied" 7 (6.14percent) are „Very Dissatisfied" in using UGC-eSS Consortium. In the case of „Faculty", 11 (12.79percent) are „Very Satisfied", 30 (34.88percent) are „Satisfied" and only 3 (3.49percent) are „Very Dissatisfied". It is identified from the table that majority 44.50per cent of the respondents are „Satisfied" the opinion in using UGC-eSS consortium.

4.29 RATING THE DIFFICULTIES ENCOUNTERED IN ACCESSING E-JOURNALS

The study has been analysed the Difficulties encountered in accessing E- journals- by the users in Madras University library. The five point scales of Never, Rarely, Sometimes, Often and More Often were used for the study. The Mean, Standard Deviation and their Rank for the Difficulties encountered in accessing E- journals have been calculated and it is shown in the table 4.28.

Table 4.28 Rating the Difficulties Encountered in Accessing E-journals

Sl. No	Description	Never	Rarely	Some-times	Often	Most Often	—	Std. Dev.	R
1	Required Password not Known	6	5	34	85	71	4.07	0.957	4
2	Access is restricted to Campus Only	6	5	18	101	71	4.23	0.879	2
3	The E-Journal Interface is difficult to Navigate	9	8	17	75	92	4.27	1.036	1
4	The Library does not have subscription to Journals required	11	8	13	85	83	4.17	1.047	3

Source: Primary Data

It is identified from Table 4.28 that the respondents are most often encountered with the difficulties in accessing E-journals like „The E-Journal Interface is difficult to Navigate", „Access is restricted to Campus Only", and „The Library does not have subscription to Journals required". The respondents are often encountered with the difficulties of „Required Password not Known". The mean value of all the variables range between 4.07 and 4.27. It can be inferred that all the four variables lie on „More Often" and often. The standard deviation of opinion ranges between 0.879 and 1.047.

4.30 DIFFICULTIES ENCOUNTERED IN ACCESSING E-JOURNALS

The study has been further extended to respondent category for the Difficulties encountered in accessing E-journals by the users in university library. The Mean, Standard Deviation and their Rank for the Difficulties encountered in accessing E- journals have been calculated and it is shown in the table 4.29.

Table 4.29 Difficulties Encountered in Accessing E-journals

Sl. No	Description	Research Scholars			Faculty			Chi- Square
		—	Std. Dev.	R	—	Std. Dev.	R	
1	Required Password not Known	4.11	0.894	4	4.20	1.111	3	33.128
2	Access is restricted to Campus Only	4.27	0.886	2	4.23	0.901	2	12.347
3	The E-Journal Interface is difficult to Navigate	4.21	0.997	3	4.33	1.037	1	24.765
4	The Library does not have subscription to Journals required	4.39	0.726	1	4.01	1.218	4	31.021

Source: Primary Data

Note : $\bar{=}$-Mean, SD-Standard Deviation, R-Rank, Degrees of Freedom: 4, Table Value –9.488.

It is identified from Table 4.29 that the respondents in the category of „Research Scholars" are more often encountered with difficulties in accessing E- journals like „The Library does not have subscription to Journals required", „Access is restricted to Campus Only" and „The E-Journal Interface is difficult to Navigate. The respondents are more often encountered with difficulties of „Required Password not Known". The mean value of all the variables range between 4.11 and 4.39. It is inferred that all the four variables lie on „More Often". The standard deviation of opinion ranges between 0.726 and 0.997.

The respondents in the category of „faculty" are more often encountered with difficulties in accessing E-journals like „The E-Journal Interface is difficult to Navigate". „Access is restricted to Campus Only" and „Required Password not

Known". The respondents are often facing the problems of „The Library does not have subscription to Journals required". The mean value of all the variables range between 4.01 and 4.33. It is inferred that all the four variables lie on „Often" and „More Often". The standard deviation of opinion ranges between 0.901 and 1.218.

Further, Chi square has been administered to identify the significance. The calculated value has been shown in Table 4.29. The table value is 9.488 at 5percent level of significance, the computed value of all the variables are higher than the table value which indicated the variables are significant in their opinion between the respondent categories towards the Difficulties encountered in accessing E-journals by the users except the variable of „Access is restricted to Campus Only".

4.31 THE LEVEL OF SATISFACTION WITH ELECTRONIC SOURCES FOR RESEARCH PURPOSE

The level of Satisfaction with Electronic sources for research purpose among the users in University libraries has been analysed based on the opinion of the respondents and it is shown in the table 4.30

Table 4.30 The level of Satisfaction with Electronic Sources for Research Purpose

Sl. No	Description	Respondents' Category		Total Respondents
		Research Scholars	Faculty	
1	Not helpful	4 (3.51)	4 (4.65)	8 (4.00)
2	No comments	23 (20.18)	15 (17.44)	38 (19.00)
3	Helpful	53 (46.49)	43 (50.00)	96 (48.00)
4	Most helpful	34 (29.82)	24 (27.91)	58 (29.00)
	Total	**114 (100.00)**	**86 (100.00)**	**200 (100.00)**

Source: Primary Data

Note : Figure in brackets is percentages to total.

It revealed from the table 4.30, that the level of satisfaction with electronic sources analysed with their respondent category . In the category of „Research Scholars", 4 (3.51percent) views „Not helpful" and 23 (20.18percent) views „No comments" with Electronic Sources for Research purpose followed by 34 (29.82percent) views „Most Helpful" towards Electronic Sources for Research purpose. In the case of faculty, 4 (4.65percent) views „Not Helpful", 15 (17.44percent) views „No comments" and 24 (27.91percent) views for „Most Helpful" towards Electronic Sources for Research purpose. It is identified from the table that majority (96) 48 per cent of the respondents has replied that the Electronic sources are „Helpful" for research purpose.

4.32 THE EXTENT OF USE OF THE ELECTRONIC RESOURCES AND PRINTED MATERIALS IN OWN WORK

The extent of use of the electronic resources and printed materials in own work among the users in University libraries has been analysed based on the opinion of the respondents and it is shown in the table 4.31

Table 4.31 The extent of use of the Electronic Resources and Printed Materials in Own Work

Sl. No	Description	Respondents Category		Total Respondents
		Research Scholars	Faculty	
1	Use Only Electronic Resources	27 (23.68)	24 (27.91)	51 (25.50
2	Use both Electronic and Printed Materials But Mostly Electronic	39 (34.21)	12 (13.95)	51 (25.50)
3	Use Electronic and Printed Materials Equally	45 (40.35)	37 (43.02)	83 (41.50)
4	Download Electronic But Convert to Print For Reading	1 (0.88)	11 (12.79)	12 (6.00)
5	Use mostly Printed Materials	1 (0.88)	2 (2.33)	3 (1.50)
	Total	**114 (100.00)**	**86 (100.00)**	**200 (100.00)**

Source: Primary Data

Note : Figure in brackets is percentages to total.

It is revealed from the table 4.31, that the extent of use of the electronic resources and printed materials in own work analysed with their respondents category. In the category of Research Scholars, 27 (23.68percent) has used „Only for Electronic Resources", 45 (40.35percent) has used for „both Electronic and

Printed Materials. In the case of faculty members 37 (43.02percent) has used for „both Electronic and Printed Materials and 24 (27.91percent) has used only Electronic Resources. It is identified from the table that majority 41.50 per cent of the respondents has used „both Electronic and Printed materials equally".

4.33 PERCEIVED ADVANTAGES OF E-RESOURCES

The study has been analysed the perceived advantages of E-resourcesby the users in university library. The five point scales of Strongly Disagree, Disagree, No Opinion, Agree, Strongly Agree were used for the study. The Mean, Standard Deviation and their Rank for the Perceived advantages of E-resources have been calculated and it is shown in the table 4.32.

Table 4.32 Perceived Advantages of E-resources

Sl.No	Description	SDA	DA	NO	A	SA	—	Std. Dev.	Rank
1	Search ability	2	17	22	74	85	4.20	0.987	6
2	Universal access	1	1	30	102	65	4.26	0.784	5
3	Easy to Handle	0	3	22	98	77	4.41	0.716	3
4	Hyperlinks	1	5	26	129	39	4.17	0.688	7
5	Data Integrity/Security	0	3	47	99	50	4.02	0.701	8
6	Highly Storage	3	6	18	43	129	4.66	0.893	1
7	Saves space/time	1	5	20	97	77	4.33	0.754	4
8	Consortia Approach	3	11	67	69	50	3.71	0.937	10
9	Multiple access/Use	2	5	32	39	122	4.61	0.914	2
10	Reduced costs	2	5	50	111	32	3.93	0.919	9

Source: Primary Data

Note : SDA- Strongly Disagree, DA- Disagree, NO- No Opinion, A- Agree, SA- Strongly Agree.

It is assessed from Table 4.32 that the respondents has strongly agreed for the advantage of „Highly Storage", „Multiple access/Use" and „Easy to Handle". The respondents has No opinion regarding the advantage of e-resources are „Consortia Approach". The mean value of all the variables range between 3.71 and 4.66. It is inferred that all the ten variables lie between „Strongly Disagree to Strongly Agree". The deviation of opinion ranges between 0.688 and 1.987

4.34 PERCEIVED ADVANTAGES OF E-RESOURCES

The study has been further extended to respondent category for the Perceived advantages of E-resources by the users in universities library. The Mean, Standard Deviation and their Rank for the advantages of Electronic information resources have been calculated and it is shown in the table 4.33.

Table 4.33 Perceived advantages of E-resources

Sl. No	Description	Research Scholar			Faculty			Chi- Square
		—	Std. Dev.	R	—	Std. Dev.	R	
1	Search ability	4.28	0.816	3	3.98	1.157	7	36.854
2	Universal access	4.07	0.687	8	4.36	0.788	3	33.128
3	Easy to Handle	4.14	0.636	6	4.56	0.793	1	27.584
4	Hyperlinks	4.09	0.699	7	3.84	0.810	8	18.495
5	Data Integrity/Se-curity	4.19	0.614	5	3.64	0.886	9	40.358
6	Highly Storage	4.47	1.023	2	4.51	0.964	2	10.875
7	Saves space/time	4.26	0.773	4	4.27	0.793	5	9.843
8	Consortia Approach	3.92	1.037	9	3.60	1.035	10	15.245
9	Multiple access/Use	4.52	0.864	1	4.30	1.087	4	18.912
10	Reduced costs	3.87	0.933	10	4.09	0.958	6	17.536

Source: Primary Data

Note : ¯-Mean, SD-Standard Deviation, R-Rank, Degrees of Freedom: 4, Table Value – 9.488.

It is analysed from Table 4.33 that the respondents in the category of „Research Scholar" has strongly agreed to the benefits of E-resources such as „Multiple access/Use, Highly Storage" and „Search ability". The respondents also agreed that they receive the ;benefit of „Reduced costs". The mean value of all the variables range between 3.87 and 4.52. It can be inferred that all the ten variables lie between „ Agree" and „Strongly Agree". The standard deviation of opinion ranges between 0.614 and 1.023.

Further, Chi square has been administered to identify the significance. The computed value has been shown in Table 4.33. The table value is 9.488 at 5percent level of significance, the calculated value of all the variables are higher than the table value which indicated the variables are significant in their opinion between the respondent categories towards the perceived advantages of electronic information resources by the users.

4.35 LEVEL OF SATISFACTION OF INFORMATION OBTAINED FROM LIBRARY USING E- RESOURCES

The Level of satisfaction of information obtained from library using E-resources among the users in University library has been analysed based on the opinion of the respondents and it is shown in the table 4.34.

Table 4.34 Level of Satisfaction of Information obtained from Library using E- resources

Sl. No	Description	Respondents Category		Total Respondents
		Research Scholars	Faculty	
1	Highly satisfied	34 (29.82)	20 (23.26)	54 (27.00)
2	Satisfied	35 (30.70)	35 (40.70)	70 (35.00)
3	No Comments	16 (14.04)	8 (9.30)	24 (12.00)
4	Partially satisfied	15 (13.16)	16 (18.60)	31 (15.50)
5	Dissatisfied	14 (12.28)	7 (8.14)	21 (10.50)
	Total	**114 (100.00)**	**86 (100.00)**	**200 (100.00)**

Source: Primary Data

Note : Figure in brackets is percentages to total.

It is revealed from the table 4.34, that the level of satisfaction of information obtained from library using E-resources with their respondents category. In the category of Research Scholars, 34 (29.82percent) are „highly satisfied", 35 (30.70percent) are „Satisfied" and 15 (13.16percent) are „partially satisfied" towards information obtained from library using E-resources. In the case of faculty, 35 (40.70percent) are „satisfied" and 20(23.26percent) are Highly Satisfied" regarding information obtained from library using E-resources. It is identified from the table that majority 35 per cent of the respondents are „Satisfied" with the information obtained from library using E-resources.

4.36 THE SATISFACTION ABOUT THE AVAILABILITY OF E-RESOURCES

The study has been analysed the satisfaction about the availability of E-resources by the users in university library. The five point scales of Poor, No comments, Satisfaction, Good, Very Good were used for the study. The Mean, Standard Deviation and their Rank for the satisfaction about the availability of E-resources have been calculated and it is shown in the table 4.35.

Table 4.35 The Satisfaction about the availability of E-resources

Sl. No	Description	Poor	No comments	Satisfaction	Good	Very good	—	Std. Dev.	Rank
1	Quality	28	44	10	70	48	3.89	1.215	6
2	Current	2	4	20	94	80	4.28	0.812	3
3	Authenticity	1	7	19	110	63	4.21	0.841	4
4	Easy to retrieve	1	14	17	47	120	4.41	1.127	1
5	Downloading speed	2	7	47	106	39	4.12	0.736	5
6	Up to date-ness	2	2	10	118	68	4.36	0.724	2

Source: Primary Data

It is identified from Table 4.35 that the respondents are rated very good about the availability of E-resources factors like „Easy to retrieve", „Up to dateness" and„Current Topics". The respondents also rate good about the factor „Quality". The mean value of all the variables range between 3.89 and 4.41. It is inferred that all the six variables lie between „Good" and „Very Good". The standard deviation of opinion ranges between 0.724 and 1.127.

4.37 THE SATISFACTION ABOUT THE AVAILABILITY OF E-RESOURCES

The study has been further extended to respondents category for the satisfaction about the availability of E-resources by the users in universities library. The Mean, Standard Deviation and their Rank for the satisfaction of availability of Electronic information resources have been calculated and it is shown in the table 4.36.

It is analysed from Table 4.36 that the respondents in the category of „Research Scholars" has viewed very good for the factor such as „Easy to retrieve", „Current Topics" and „Up to datedness" The least preference was given for the type of „Quality". The mean value of all the variables range between 3.68 and 4.38. It can be inferred that all the six variables lie between „Good" and „Very Good". The deviation of opinion ranges between 0.822 and 1.218.

Table 4.36: The satisfaction about the availability of E-resources

Sl. No	Description	Research Scholar			Faculty			Chi-Square
		$\bar{x}$	Std. Dev.	R	$\bar{x}$	Std. Dev.	R	
1	Quality	3.68	1.218	6	3.78	1.356	6	14.659
2	Current	4.38	0.877	2	4.24	0.858	3	11.347
3	Authenticity	4.21	0.822	4	4.12	0.816	4	7.124
4	Easy to retrieve	4.47	1.120	1	4.39	0.977	1	3.621
5	Downloading speed	4.09	0.983	5	3.91	0.779	5	22.781
6	Up to datedness	4.26	0.849	3	4.31	0.635	2	9.109

Source: Primary Data

Note : $\bar{x}$-Mean, SD-Standard Deviation, R-Rank, Degrees of Freedom: 4, Table Value – 9.488.

The respondents in the category of „faculty" has opinion of very good towards the availability of E-resources factors like „Easy to retrieve", „Up to datedness"and „Current topics". The respondents feels good towards „Quality". The mean value of all the variables range between 3.78 and 4.39. It is inferred that all the six variables lie between „Good" and „Very Good". The standard deviation of opinion ranges between 0.635 and 1.356.

Further, Chi square has been administered to identify the significance. The computed value has been shown in Table 4.36. The table value is 9.488 at 5percent level of significance, the calculated value for the variables of „Quality", „Current" and„ Downloading speed" are higher than the table value which indicated the variables are significant in their opinion between the respondent categories towards the satisfaction about the availability of by the users except the variables of „Authenticity", „Easy to retrieve" and „Up to datedness".

4.38 RATING THE IMPORTANCE OF USING ELECTRONIC RESOURCES

The study has been analysed the importance of using electronic resources by the users in universities library. The five point scales of Not important at all, Not important, Somewhat important, Important and Very important were used for the study. The Mean, Standard Deviation and their Rank for the importance of using E-resources have been calculated and it is shown in the table 4.37.

It is studied from Table 4.37 that the respondents are giving very much importance towards usage of Downloading programs", „Placements" and „Accessing online databases". The respondents are giving importance to usage of „Research/Project". The mean value of all the variables range between 3.63 and 4.67. It can be inferred that all the eight variables lie between „Important" and „Very Important". The deviation of opinion ranges between 0.843 and 1.128.

Table 4.37: Rating the importance of using Electronic Resources

Sl. No	Description	Not Important at all	Not Important	Somewhat Important	Important	Very Important	—	Std. Dev.	R
1	Research/ Project	2	36	91	10	60	3.63	1.128	8
2	Lesson Plans	1	2	52	69	77	4.17	0.847	5
3	Accessing online databases	1	2	47	60	90	4.31	0.864	3
4	Publishing	1	29	32	42	97	4.09	1.117	6
5	Downloading programs	1	7	19	27	146	4.67	0.843	1
6	Placements	2	5	32	23	137	4.58	0.927	2
7	Chatting	3	20	22	110	46	3.87	0.922	7
8	Entertainment	3	3	21	115	57	4.24	0.766	4

Source: Primary Data

4.39 THE IMPORTANCE OF USING ELECTRONIC RESOURCES

The study has been further extended to respondents category for the importance of using the Electronic information resources by the users in universities library. The Mean, Standard Deviation and their Rank for the importance of using the Electronic information resources have been calculated and it is shown in the table 4.38.

Table 4.38 The Importance of Using Electronic Resources

Sl. No	Description	Research Scholars			Faculty			Chi-Square
		–	Std. Dev.	R	–	Std. Dev.	R	
1	Research/Project	3.87	1.057	8	3.79	1.154	8	27.459
2	Lesson Plans	4.41	0.918	4	4.08	0.866	6	6.214
3	Accessing online databases	4.47	0.837	3	4.18	0.887	4	12.417
4	Publishing	4.19	1.212	6	4.11	1.078	5	13.124
5	Downloading programs	4.67	0.917	1	4.58	1.019	1	7.514
6	Placements	4.59	0.878	2	4.41	1.037	2	18.254
7	Chatting	4.09	1.010	7	4.01	0.943	7	20.271
8	Entertainment	4.24	0.754	5	4.22	0.815	3	14.224

Source: Primary Data

Note :¯-Mean, SD-Standard Deviation, R-Rank, Degrees of Freedom: 4, Table Value – 9.488.

It is inferred from Table 4.38 that the respondents in the category of „Research Scholars" are giving very much importance to „Downloading programs", „Placements" and „Accessing online databases". The respondents are giving importance of „Research/Project". The mean value of all the variables range between 3.87 and 4.67. It is inferred that all the eight variables lie between„ Important" and „Very Important". The standard deviation of opinion ranges between 0.754 and 1.212.

Similarly, the respondents in the category of „Faculty" gives very much importance to „Downloading programs", „Placements" and „Entertainment". The respondents are giving importance to „Research/Project". The mean value of all the variables ranges between 3.79 and 4.58. It is inferred that all the eight variables lies between „Important" and „Very Important". The standard deviation of opinion ranges between 0.815 and 1.154.

Further, Chi square has been administered to identify the significance. The computed value has been shown in Table 4.38. The table value is 9.488 at 5percent level of significance, the computed value of most of the variables were higher than the table value which indicated the variables are significant in their opinion between the respondent categories towards the purpose of using the Electronic information resources by the users except the variables of „Lesson Plans" and „Downloading Programs".

4.40 THE LEVEL OF SATISFACTION OF THE ELECTRONIC RESOURCES

The study has been analysed the level of Satisfaction of the electronic resources accessed by the users in universities library. The five point scales of Strongly Disagree, Disagree, No Opinion, Agree, Strongly Agree are used for the study. The Mean, Standard Deviation and their Rank for the level of Satisfaction of the electronic resources have been calculated and it is shown in the table 4.39.

It is released from Table 4.39 that the respondents are strongly agrees to the satisfaction of „Library provides adequate electronic resources" and „Library provides adequate training on how to use the electronic resources". The respondents also strongly agreed with the type of „Electronic information develops academic skills". The mean value of all the variables range between 4.12 and 4.69. It is inferred that all the four variables lie on „Strongly Agree". The standard deviation of opinion ranges between 0.8185 and 1.039.

Table 4.39 The Level of satisfaction of the Electronic Resources

Sl.No	Description	SDA	DA	NO	A	SA	—	Std. dev.	R
1	Library provides adequate training on how to use the electronic resources	2	3	10	125	60	4.23	0.875	3
2	Electronic information develops academic skills	2	25	20	71	82	4.12	1.039	4

Sl.No	Description	SDA	DA	NO	A	SA	—	Std. dev.	R
3	Library provides adequate electronic resources	1	7	19	73	100	4.56	0.818	2
4	level of satisfaction while using electronic information resources	1	4	25	31	139	4.69	0.854	1

Source: Primary Data

Note: SDA- Strongly Disagree, DA- Disagree, NO- No Opinion, A- Agree, SA- Strongly Agree.

4.41 THE LEVEL OF SATISFACTION OF THE ELECTRONIC RESOURCES

The study has been further extended to respondent category for the Level of satisfaction of the Electronic information resources by the users in universities library. The Mean, Standard Deviation and their Rank for the Level of satisfaction of the Electronic information resources have been calculated and it is shown in the table 4.40.

Table 4.40 The Level of Satisfaction of the Electronic Resources

Sl. No	Description	Research scholar			Faculty			Chi-Square
		—	Std. Dev.	R	—	Std. Dev.	R	
1	Library provides adequate training on how to use the electronic resources	4.39	0.742	3	4.48	0.739	3	14.657
2	Electronic information develops academic skills	4.31	1.024	4	4.01	1.131	4	13.610
3	Library provides adequate electronic resources	4.41	0.848	2	4.54	0.923	2	9.744
4	level of satisfaction while using electronic information resources	4.58	0.914	1	4.62	0.813	1	2.514

Source: Primary Data

Note : ‾-Mean, SD-Standard Deviation, R-Rank, Degrees of Freedom: 4, Table Value – 9.488.

It is evident from Table 4.40 that the respondents in the category of „Research Scholars" has strongly agreed to the „level of satisfaction while using electronic information resources". „Library provides adequate electronic resources" and „Library provides adequate training on how to use the electronic resources". The respondents agreed that they are satisfied with the Electronic information develops academic skills". The mean value of all the variables range between 4.31 and 4.58. It is inferred that all the four variables lie on „Strongly Agree and Agree". The deviation of opinion ranges between 0.848 and 1.024.

Simultaneously, the respondents in the category of „faculty" are strongly agreed towards „level of satisfaction while using electronic information resources". „Library provides adequate electronic resources" and „Library provides adequate training on how to use the electronic resources". The respondents are agreed towards satisfaction of Electronic information develops academic skills". The mean value of all the variables range between 4.01 and 4.62. It is inferred that all the four variables lie between „Agree" and „Strongly Agree". The standard deviation of opinion range between 0.739 and 1.131.

Further, Chi square has been administered to identify the significance. The computed value has been shown in Table 4.40. The table value is 9.488 at 5% level of significance, the computed value of most of the variables are higher than the table value which indicated the variables are significant in their opinion between the respondent categories towards the level of satisfaction of the Electronic information resources by the users except the variable of „level of satisfaction while using electronic information resources".

4.42 PURPOSES OF USING ELECTRONIC INFORMATION SERVICES

The study has been analysed the Purposes of using electronic Information Services by the users in universities library. The five point scales of Strongly Disagree, Disagree, No Opinion, Agree, Strongly Agree were used for the study. The Mean, Standard Deviation and their Rank for the purposes of using have been calculated and it is shown in the table 4.41.

Table 4.41 Rating the Purposes of using electronic Information Services

Sl. No	Description	SDA	DA	NO	A	SA	—	Std. Dev.	R
1	To know the availability of materials	2	3	18	15	162	4.74	0.818	1
2	To get study/research related materials	6	13	12	89	79	4.28	1.027	3

Sl. No	Description	SDA	DA	NO	A	SA	—	Std. Dev.	R
3	To get material for class assignments	6	13	45	57	79	3.95	1.121	5
4	To get literature for research	7	8	32	63	90	4.14	1.066	4
5	To get the print out of required materials	6	7	34	44	108	4.57	1.038	2

Source: Primary Data

Note : SDA- Strongly Disagree, DA- Disagree, NO- No Opinion, A- Agree, SA- Strongly Agree.

It is revealed from Table 4.41 that the respondents are strongly agreed for the purpose of using Electronic Information Services „To know the availability of materials", „To get the print out of required materials" and „To get study/research related materials". The respondents no opinion about the purpose of using Electronic Information Services. To get material for class assignments". The mean value of all the variables range between 3.95 and 4.74. It is inferred that all the five variables lie between „No opinion to Strongly Agree". The standard deviation of opinion range between 0.818 and 1.121. 4.41.Purposes of using electronic Information Services

The study has been further extended to respondent category for the purpose of using the Electronic information services by the users in universities library. The Mean, Standard Deviation and their Rank for the purpose of using the Electronic information services have been calculated and it is shown in the table 4.42.

Table 4.42 Purposes of using Electronic Information Services

Sl. No	Description	Research scholar			Faculty			Chi-Square
		—	Std. Dev.	R	—	Std. Dev.	R	
1	To know the availability of materials	4.84	0.877	1	4.69	0.809	1	15.839
2	To get study/research related materials	4.63	1.068	2	4.16	0.826	5	18.612
3	To get material for class assignments	3.87	1.038	5	4.39	1.063	4	26.123
4	To get literature for research	4.08	1.082	4	4.48	1.077	3	27.219

Sl. No	Description	Research scholar			Faculty			Chi-Square
		—	Std. Dev.	R	—	Std. Dev.	R	
5	To get the print out of required materials	4.24	1.033	3	4.55	1.021	2	12.327

Source: Primary Data

Note : ̄-Mean, SD-Standard Deviation, R-Rank, Degrees of Freedom: 4, Table Value – 9.488.

It is identified from Table 4.42 that the respondents in the category of „Research Scholars" has strongly agreed using Electronic Information Services in order „To know the availability of materials". „To get study/research related materials" and „To get the print out of required materials". The respondents has no opinion towards the purpose of using Electronic Information Services „To get material for class assignments". The mean value of all the variables range between 3.87 and 4.84. It is inferred that all the five variables lie between „Strongly Disagree to Strongly Agree". The standard deviation of opinion ranges between 0.877 and 1.082.

Similarly, the respondents in the category of „Faculty" has strongly agreed for purpose of using Electronic Information Services To know the availability of materials", „To get the print out of required materials" and „To get literature for research". The respondents agreed for the purpose of using „To get study/research related materials". The mean value of all the variables range between 4.16 and 4.69. It can be inferred that all the five variables lie between „Agree" and „Strongly Agree". The deviation of opinion ranges between 0.809 and 1.077.

Further, Chi square has been administered to identify the significance. The computed value has been shown in Table 4.42. The table value is 9.488 at 5% level of significance, the computed value of all the variables are higher than the table value which indicated the variables are significant in their opinion between the respondent categories towards the problems of facing in accessing the Electronic information services by the users.

4.43 THE PROBLEMS FACING IN ACCESSING ELECTRONIC INFORMATION SERVICES

The study has been analysed the Problems facing in accessing electronic information services by the users in universities library. The five point scales of Strongly Disagree, Disagree, No Opinion, Agree, Strongly Agree were used for the study. The Mean, Standard Deviation and their Rank for the problems have been calculated and it is shown in the table 4.43.

Table 4.43 Rating the Problems facing in Accessing Electronic Information Services

Sl. No	Description	SDA	DA	NO	A	SA	—	Std.Dev.	R
1	Time Consuming	47	22	16	67	49	3.58	1.424	6
2	Slow Speed	6	13	10	53	118	4.41	1.068	1
3	Not easy to search	6	13	12	54	115	4.36	1.111	2
4	Not familiar with the systems	6	13	6	110	65	4.09	0.971	4
5	Not enough terminals to access information	6	13	12	64	104	4.28	1.047	3
6	Lack of trained staff	14	41	17	73	55	3.57	1.218	5

Source: Primary Data

Note : SDA- Strongly Disagree, DA- Disagree, NO- No Opinion, A- Agree, SA- Strongly Agree..

It is revealed from Table 4.43 that the respondents strongly agreed towards the problems faced for the „Slow Speed", „Not easy to search" and „Not enough terminals to access information". The respondents has agreed towards the problems faced by „Lack of trained staff". The mean value of all the variables range between 3.58 and 4.41. It is inferred that all the six variables lie between „Agree" and „Strongly Agree". The standard deviation of opinion ranges between 0.971 and 1.218.

4.44 THE PROBLEMS FACING IN ACCESSING ELECTRONIC INFORMATION SERVICES

The study has been further extended to respondent category for the Problems facing in accessing electronic information services by the users in universities library. The Mean, Standard Deviation and their Rank for the Problems facing in accessing electronic information services have been calculated and it is shown in the table 4.44.

Table 4.44 The Problems facing in Accessing Electronic Information Services

Sl. No	Description	Research scholar			Faculty			Chi-Square
		—	Std. Dev.	R	—	Std. Dev.	R	
1	Time Consuming	3.79	1.218	6	3.67	1.350	6	17.254
2	Slow Speed	4.36	1.054	2	4.66	1.047	1	15.645

Sl. No	Description	Research scholar			Faculty			Chi-Square
		—	Std. Dev.	R	—	Std. Dev.	R	
3	Not easy to search	4.29	1.108	3	4.51	1.157	2	18.387
4	Not familiar with the systems	4.11	0.919	4	4.31	0.991	3	29.468
5	Not enough terminals to access information	4.47	1.121	1	4.24	0.944	4	23.158
6	Lack of trained staff	3.34	1.218	5	4.09	1.119	5	14.282

Source: Primary Data

Note : ¯-Mean, SD-Standard Deviation, R-Rank, Degrees of Freedom: 4, Table Value – 9.488.

It is identified from Table 4.44 that the respondents in the category of „Research scholars" Strongly agreed towards the problems of „Not enough terminals to access information", „Slow speed" and „Not easy to search". The respondents agreed towards the problems of „Time Consuming". The mean value of all the variables range between 3.79 and 4.47. It is inferred that all the six variables lie between „Agree" and „Strongly Agree". The standard deviation of opinion ranges between 0.919 and 1.218.

The category of „Faculty", the respondents strongly agreed to the problem of „Slow speed", „Not easy to search" and „Not familiar with the systems". The respondents agreed to the problems of „Time Consuming". The mean value of all the variables ranges between 3.67 and 4.66. It is inferred that all the six variables lies between „Agree" and „Strongly Agree". The deviation of opinion ranges between 0.944 and 1.350.

Further, Chi square has been administered to identify the significance and the hypothesis "There is no significance difference between the respondent categories of research scholars and faculties towards the problems facing in accessing electronic information services". The computed value has been shown in Table 4.43. The table value is 9.488 at 5% level of significance, the computed value of all the variables are higher than the table value which indicated the variables are significant in their opinion between the respondent categories towards the problems facing in accessing electronic information services by the users. Therefore the third hypothesis namely "There is no significance difference between the respondent categories of research scholars and faculties towards the problems facing in accessing electronic information services is invalid and rejected.

4.45 VALUE AND IMPORTANCE OF THE ELECTRONIC REFERENCE SERVICES

The study has been analysed the Value/ importance of the electronic reference services by the users in University library. The five point scales of Not important at all, Not important, Somewhat important, Important, Very Important were used for the study. The Mean, Standard Deviation and their Rank for the Value/ importance of the electronic reference service have been calculated and it is shown in the table 4.45.

Table 4.45 Rating the Value and Importance of the Electronic Reference Services

Sl. No	Description	Not Importan-tat all	Not Important	Somewhat Important	Important	Very Important	–	Std. Dev	R
1	E-mail Reference	6	13	12	62	106	4.37	1.040	1
2	Web forms	1	6	20	111	61	4.20	0.765	2
3	Ask-A Librarian Service	1	5	85	79	30	3.87	0.794	5
4	Online Chat Reference	1	5	50	85	59	3.99	0.869	4
5	Social Media	1	5	32	101	61	4.14	0.844	3

Source: Primary Data

Note : M-Mean, SD-Standard Deviation, R-Rank, Degrees of Freedom: 4, Table Value – 9.488.

It is revealed from Table 4.45 that the respondents consider electronic reference services of „E-mail Reference", „Web forms" and „Social Media" are to be as very important. The respondents consider electronic reference services of „Ask-A Librarian Service" as an important one. The mean value of all the variables range between 3.87 and 4.37. It is inferred that all the five variables lie between „Important" and „Very Important". The standard deviation of opinion ranges between 0.765 and 1.040.

4.46 VALUE AND IMPORTANCE OF THE ELECTRONIC REFERENCE SERVICES

The study has been further extended to respondents category for the Value/ importance of the electronic reference services by the users in university library. The Mean, Standard Deviation and their Rank for the Value/ importance of the electronic reference services have been calculated and it is shown in the table 4.46.

Table 4.46 Value and importance of the Electronic Reference Services

Sl. No	Description	Research scholar			Faculty			Chi- Square
		-	Std. Dev.	R	-	Std. Dev.	R	
1	E-mail Reference	4.37	1.087	1	4.48	0.947	1	18.548
2	Web forms	4.21	0.759	2	4.27	0.814	3	6.863
3	Ask-A Librarian Service	4.01	0.778	5	3.89	0.776	5	28.197
4	Online Chat Reference	4.18	0.768	3	4.18	0.966	4	15.337
5	Social Media	4.11	0.843	4	4.33	0.917	2	18.216

Source: Primary Data

Note :¯-Mean, SD-Standard Deviation, R-Rank, Degrees of Freedom: 4, Table Value – 9.488.

It is analysed from Table 4.46 that the respondents in the category of „Research Scholars" has consider the electronic reference services of „E-mail Reference". „web forms" and „Online Chat Reference" as very important. The respondents considered electronic reference service like „Ask-A Librarian Service" as an important one. The mean value of all the variables range between 4.01 and 4.37. It can be inferred that all the five variables lie between „Important" and „Very Important". The standard deviation of opinion ranges between 0.759 and 1.087.

The respondents in the category of „Faculty" has considered the electronic reference services of „E-mail Reference", „Social Media" and „Web forms" as very important. The respondents has considered Ask-A Librarian Service" as an important. The mean value of all the variables range between 3.89 and 4.48. It can be inferred that all the five variables lie between „Important" and „Very Important". The standard deviation of opinion ranges between 0.776 and 0.947.

Further, Chi square has been administered to identify the significance. The computed value has been shown in Table 4.45. The table value is 9.488 at 5% level of significance, the computed value of most of the variables are higher than the table value which indicated the variables are significant in their opinion between the respondent categories towards the Value/ importance of the electronic reference services by the users except the variable of „Web forms".

4.47 SUMMARY

The study reveals that, that majority 35.53 per cent of the respondents are spending in library about 1 to 2 hours per day. Nearly 35 per cent of the respondents are„ Satisfied" with the information obtained from library using E-resources. The study further reveals that, that around 39.64 per cent of the respondents are accessing the electronic information from Madras University central library.

The study reveals that the respondents in the category of „Research Scholars" are highly satisfied with „Easy to retrieve", „Current Topics" and „Up to datedness". The respondents are satisfied with „Quality" of library service. The major findings are extracted in the chapter after analysing the data and interpretations with suggestions and conclusion.

Chapter - V

Findings, Suggestions and Conclusion

5.1 INTRODUCTION

The previous chapter, presented a detailed discussion on the analysis of the data collected for "Information Literacy in Madras University Library, Chennai.. In this chapter, the findings, suggestions and conclusions based on the outcomes of the study are highlighted.

5.2 FINDINGS

1. The study reveals that, the gender-wise distribution of respondents to use in Madras University Library. From the 200 respondents, 59.50 per cent are belongs to male and the remaining 40.50 per cent are female.

2. Out of 275 respondents, 200 replied after duly filled by the respondents which comprises 119 (59.50%) were from male respondents and 81 (40.50%) were from female respondents. The response rate is 72.73 per cent.

3. Further the respondents were classified under their respondent category. Among the 200, 114 (57.00%) of the respondents are scholars and the remaining 86 (43.00%) of them were faculty members.. It is concluded from the table that majority 57.00 per cent of the respondents were in the age category of research scholars.

4. The respondents were classified under their age frequency taken for the study. From the 200, 38 (19.00%) were distributed to the frequency of „up to 30", 43 (21.50%) were distributed to the frequency of „31-35" years, 43 (21.50%) were distributed to the frequency of „36-40" years of age and 76 (38.00%) were distributed to the frequency of „above 40" years of age. Majority of the respondents are above 40 years of age group.

5. The study shows that majority of the respondents 67 (33.47%) visit the library twice a week.

6. The study shows that the frequency of time spend in the library among the users in the Madras university library with their respondent category. From the category of research scholars, 48 (42.16%) are spending 1 to 2 hours per day, 25 (21.96%) are spending 2 to 3 hours per day and 19 (16.69%) are spending 30 minutes to 1 hour. In the case of faculty members, 32 (37.21%) of the respondents are spending 30 minutes to one hour, 24 (27.91%) are spending upto 30 minutes and 23 (25.58%) were spending 1 to 2hours. It is identified from the table that majority 35.03 per cent of the respondents are spending library in 1to 2 hours per day.

7. It is identified the study that the respondents have given first priority for the purpose of „E-resources and others". „general knowledge and employment news" and „entertainment" are the second and third preference given by the respondents. The least preference was given for the purpose of „academic improvement and research purpose". The mean value of all the variables range between 3.18 and 4.36. It can be inferred that all the seven variables lie between „often and „more often". The standard deviation of opinion ranges between 0.879 and 1.286.

8. The chi square has been administered to identify the significance and the hypothesis "There is no significance difference between the respondent categories of research scholars and faculties towards the purpose of using library". The table value is 9.488 at 5% level of significance, the calculated value of all the variables are higher than the table value which indicated the variables are significant in their opinion between the respondent categories of research scholars and faculties towards the purpose of using libraries. Therefore the first hypothesis namely "There is no significance difference between the respondent categories of research scholars and faculties towards the purpose of using library" is invalid and rejected.

9. Majority 32.82 per cent of the respondents are „agreed" the opinion about the development of general knowledge after going to the Library

10. It is identified from the stduy that majority 39.64 per cent of the respondents are accessed the electronic information from Madras university central library.

11. The study shows that the level of using the electronic information resources among the users in the university library analysed with their respondents category. From the category of research scholars, 10 (9.05%) are in the level of beginner, 65 (56.85%) are in the level of average and the remaining 39 (31.11%) of the respondents are in the level of expert. In the case of faculty

member, 5 (5.37%) are in the level of beginner and 41 (47.90%) are in the level of average and the remaining 40 (46.73%) of the respondents are in the level of expert. It is identified from the table the majority of the respondents 106 (53.02%) of the respondents are average in the level of using the electronic information resources.

12. The study reveals that majority of the respondents 102 (50.89%) are in the category of research scholars and they are frequently used the electronic information resources.
13. Majority 45.09 per cent of the respondents are spending more than 3 hours per day for the electronic information.
14. The study discloses that the methods to learn the use of E- resources among the users in the university library. Among the research scholars, 27 (23.68%) respondents are learning by attending courses, 24 (21.05%) are learning from formal training and another 24 (21.05%) are learning from teachers.. In the case of faculty members 31 (36.05%) respondents are learning from teachers, and 18 (20.93%) are learning from trail & error method and self instruction. It is identified from the table the majority 27.50 per cent of the respondents are learning from teachers to the use of E- resources.
15. It is inferred from the study that the respondents have given first priority for the type of „E-mail". „Search engines" and „Websites" are the second and third preference given by the respondents. The least preference is given for the type of „Bulletin board services". The mean value of all the variables range between 3.63 and 4.68. It can be inferred that all the eight variables lie between „Often and „More Often". The standard deviation of opinion ranges between 0.847 and 1.104.
16. The study shows that chi square has been administered to identify the significance. The table value is 9.488 at 5% level of significance, the calculated value is higher than the table value which indicated the variables are significant in their opinion between the respondent categories towards the types of E-resources accessed by the users.
17. The study reveals that the respondents have given first priority for the purpose of „To keep abreast with the latest development". „For teaching preparation" and „For career development" are the second and third preference given by the respondents. The least preference is given for the type of „E-shopping". The mean value of all the variables range between 3.78 and 4.48. It is inferred that all the nine variables lie between „twice in a week" and „daily". The standard deviation of opinion ranges between 0.874 and 1.114.

18. Chi square has been administered to identify the significance. The table value is 9.488 at 5% level of significance, the calculated value of most of the variables are higher than the table value which indicated the variables are significant in their opinion between the respondent categories towards the purpose of using the electronic information resources by the users except the variables of „To keep abreast with the latest development" (χ^2 =5.127) and „For teaching preparation (χ^2 =4.156)

19. The study found that the respondents have given first priority for the feature of „Availability", „Ease of Use" and „Flexibility" are the second and third preference given by the respondents. The least preference is given for the type of „Reliability". The mean value of all the variables range between 4.02 and 4.68. It is inferred that all the nine variables lie on „Very Good". The standard deviation of opinion ranges between 0.903 and 1.122.

20. Chi square has been administered to identify the significance. The table value is 9.488 at 5% level of significance, the calculated value of all the variables are higher than the table value which indicated the variables are significant in their opinion between the respondent categories towards the features of electronic information by the users.

21. The study shows that the acceptance of accessing information in a digital library among the users in the university library with their respondents. From the category of research scholars, 18 (15.79%) are accepting „saves more physical space", 3 (2.63%) are „library services shift from 'just in case' to 'just in time', 15 (13.16%) are „availability of information to all users simultaneously" and 29(25.44%) are for„easy retrieval".

22. The study discloses that the respondents have given first priority for the e-resource of „Internet Sources", „E- Journals" and „Open source Databases" are the second and third preference given by the respondents. The least preference is given for the type of „CD-ROM Databases". The mean value of all the variables range between 3.54 and 4.58. It is inferred that all the nine variables lie between „Satisfied" and „ Highly Satisfied". The standard deviation of opinion ranges between 0.906 and 1.384.

23. The study shows that chi square has been administered to identify the significance and the hypothesis "There is no significance difference between the respondent categories of research scholars and faculties towards the extent of satisfaction with electronic information resources".. The table value is 9.488 at 5% level of significance, the calculated value of all the variables are higher than the table value which indicated the variables are significant in

their opinion between the respondents categories towards the extent of satisfaction with electronic information resources by the users. Therefore the second hypothesis namely "There is no significance difference between the respondent categories of research scholars and faculties towards the extent of satisfaction with electronic information resources" is invalid and rejected.

24. The study shows that the accessibility of UGC- eSS E- Journals among the users in the university library with respondents. It found that 86 per cent of the respondents are more awareness and also the accessed the UGC eSS E-journals in Madras University library.

25. Majority 23 per cent of the respondents are using for the purpose of „Keeping up-to- date information".

26. It is revealed from the study that the respondents have given first priority for the problem of „Slow Internet Bandwidth". „Lack of sufficient nodes in university" and „Content searching is difficult as compared to search Engine" are the second and third preference given by the respondents. The least preference was given for the type of „Technical Problems (Server Down)".

27. Chi square has been administered to identify the significance. The table value is 15.507 at 5% level of significance, the calculated value of all the variables are higher than the table value which indicated the variables are significant in their opinion between the faculty categories towards the problems faced in using UGC- eSS E- Journals by the users in university library.

28. The study shows that the level of satisfaction in using UGC-eSS consortium among the users in the university library with their respondent category. From the category of „Research scholars", 26 (22.81%) are „Very Satisfied", 59 (51.75%) are „Satisfied" and only 7 (6.14%) are „Very Dissatisfied". In the case of „Faculty", 11 (12.79%) are „Very Satisfied", 30 (34.88%) are „Satisfied" and only 3 (3.49%) are „Very Dissatisfied". It is identified from the table that majority 44.50 per cent of the respondents are „Satisfied" the opinion inusing UGC-eSS consortium.

29. Chi square has been administered to identify the significance. The table value is 9.488 at 5% level of significance, the calculated value of all the variables are higher than the table value which indicated the variables are significant in their opinion between the respondent categories towards the Difficulties encountered in accessing E-journals by the users except the variable of „Access is restricted to Campus Only".

30. Majority 48 per cent of the respondents are replied that the Electronic sources „Helpful" for research purpose.

31. The study reveals that the extent of use of the electronic resources and printed materials in own work analysed with their respondents category. In the category of research scholars, 27 (23.68%) are used „Only Electronic Resources", 45 (40.35%) are used „both Electronic and Printed Materials. In the case of faculty members 37 (43.02%) are used „both Electronic and Printed Materials and 24 (27.91%) are used only electronic resources. It is identified from the table that majority 41.50 per cent of the respondents are used „both electronic and printed materials equally.
32. Chi square has been administered to identify the significance. The table value is 9.488 at 5% level of significance, the calculated value of all the variables are higher than the table value which indicated the variables are significant in their opinion between the respondent categories towards the perceived advantages of electronic information resources by the users.
33. Majority 35 per cent of the respondents are „Satisfied" with the information obtained from library using E-resources.
34. The study reveals that the respondents have given first priority for the satisfaction of „Easy to retrieve". „Up to dateness" and „Current" are the second and third preference given by the respondents. The least preference is given for the type of „Quality". The mean value of all the variables range between 3.89 and 4.41. It is inferred that all the six variables lie between „Good" and „Very Good". The standard deviation of opinion ranges between 0.724 and 1.127.
35. Further, Chi square has been administered to identify the significance. The table value is 9.488 at 5% level of significance, the calculated value for the variables of „Quality", „Current" and „Downloading speed" are higher than the table value which indicated the variables are significant in their opinion between the respondent categories towards the satisfaction about the availability of by the users except the variables of „Authenticity", „Easy to retrieve" and „Up to datedness".
36. The study shows that chi square has been administered to identify the significance. The table value is 9.488 at 5% level of significance, the calculated value of most of the variables are higher than the table value which indicated the variables are significant in their opinion between the respondent categories towards the purpose of using the Electronic information resources by the users except the variables of „Lesson Plans" and „Downloading Programs".
37. The study discloses that the respondents have given first priority for „level of satisfaction while using electronic information resources". „Library provides adequate electronic resources" and „Library provides adequate training

on how to use the electronic resources" are the second and third preference given by the respondents. The least preference is given for the type of „Electronic information develops academic skills". The mean value of all the variables range between 4.12 and 4.69. It is inferred that all the four variables lie on „Strongly Agree". The standard deviation of opinion ranges between 0.818 and 1.039.

38. Further, Chi square has been administered to identify the significance. The calculated value has been shown in Table 4.39. The table value is 9.488 at 5% level of significance, the calculated value of most of the variables are higher than the table value which indicated the variables are significant in their opinion between the respondent categories towards the level of satisfaction of the Electronic information resources by the users except the variable of „level of satisfaction while using electronic information resources".

39. The study discloses that the respondents have given first priority for the purpose of „To know the availability of materials". „To get the print out of required materials" and „To get study/research related materials" are the second and third preference given by the respondents. The least preference is given for the type of To get material for class assignments". The mean value of allthe variables range between 3.95 and 4.74. It is inferred that all the five variables lie between „Agree" and „Strongly Agree". The standard deviation of opinion ranges between 0.818 and 1.121.

40. Chi square has been administered to identify the significance. The table value is 9.488 at 5% level of significance, the calculated value of all the variables are higher than the table value which indicated the variables are significant in their opinion between the respondent categories towards the problems of facing in accessing the Electronic information services by the users.

41. The study reveals that the chi square has been administered to identify the significance and the hypothesis "There is no significance difference between the respondent categories of research scholars and faculties towards the problems facing in accessing electronic information services". The calculated value has been shown in Table 4.43. The table value is 9.488 at 5% level of significance, the calculated value of all the variables are higher than the table value which indicated the variables are significant in their opinion between the respondent categories towards the problems facing in accessing electronic information services by the users. Therefore the third hypothesis namely "There is no significance difference between the respondent categories of research scholars and faculties towards the problems facing in accessing electronic information services"." is invalid and rejected.

42. The study shows that the respondents have given first priority for the electronic reference services of „E-mail Reference". „Web forms" and „Social Media" are the second and third preferences given by the respondents. The least preference is given for the type of „Ask-A Librarian Service". The mean value of all the variables range between 3.87and 4.37. It is inferred that all the five variables lie between „Important" and „Very Important". The standard deviation of opinion ranges between 0.765 and 1.040.

43. Chi square has been administered to identify the significance. The calculated value has been shown in Table 4.45. The table value is 9.488 at 5% level of significance, the calculated value of most of the variables are higher than the table value which indicated the variables are significant in their opinion between the respondent categories towards the Value/ importance of the electronic reference services by the users except the variable of „Web forms".

5.3 SUGGESTIONS

- The respondents from Madras University require latest collection for academics and research.
- The electronic resources as well as the digital resources are more important for the academic and research activities. Hence, efforts should be made to develop and extract the access modes to the end users.
- Presently all are living is a digital and paperless society. Hence there is a need to increase the web based services in library activities.
- The awareness and satisfaction of library e-resources and e-services facilities are more satisfactory.
- The university should concentrate on resources as well as services like universities in Western Countries. So, the authority of institutions should provide more priorities to develop infrastructural facilities in such a library.

5.4 CONCLUSION

Presently the library plays a vital role in the academic and research field in the higher educational systems. The application of information and communication technology in libraries has greatly influenced the teaching and research community to access information, resources and services. Information and communication technology literacy skills are necessary for Faculty in the emerging knowledge driven society despite the high dissemination rate of ICT and exponential growth of Internet

The initial part of the library visits consists of frequency of library visit and time spent by the respondents. Similarly Information Access Pattern comprises of three factors such as use of E-Resources and Preferences of E-Resources and Preferred documents.

Library environment comprises of three factors such as Collections, Infrastructures and Opinion & satisfaction. Library Resources and Services comprise of three factors such as traditional resources, Digital/E-Resources and Web Resources, Satisfaction environment level comprises of three factors such as Library Facilities, Services and Professionals support. Opinion Environment comprises the Problems faced by the users and their suggestions.

The user satisfaction is the most important in measuring the quantitative and qualitative strategy to each and every library e-resources and e-services. A academic libraries are providing resources and services using technologies and tools to maximise the user satisfaction. In these circumstances, library professionals must extract e- resources and e-services among the end users. E-Resources and digitization techniques and other resources by e- services to attain the entire satisfaction of academic and research communities.

5.5 SCOPE FOR FURTHER RESEARCH

The findings of the present study gives scope for further research:

- A study may be conducted on the use of ICT infrastructure facilities available in Deemed Universities in Tamil Nadu.
- A study may be conducted on the use of E-Resources, Institutional Repository and Web technology in libraries of all affiliated colleges in University.
- A study may be conducted to develop and share the information resources and services among the students in the higher educational institutions.

Bibliography

1. Books

1. Barry M Leiner, "The Scope of the Digital Library, January 16, 1998.
2. Maharg, Paul and Paliwala, Abdul (2015). Negotiating the learning process with electronic resources. In: Effective learning and teaching in law. Kogan London, UK, 81-104.
3. Marchionini, G.Research and development in digital libraries, Allen Kent (Ed). Encyclopedia of Library and Information Science, 63: 259-279, 1998.

2. Journals

4. Agarwal, Umesh Kumar and Dave, Rajesh Kumar. (2009). Use of Internet by the Scientists of CAZRI: A survey. Indian Journal of Library and Information Science, 3(1), 15-21.
5. Ajuwon G. A. (2003). Computer and Internet Use by First Year Clinical and Nursing students in a Nigerian Teaching Hospital. BMC Medical Informatics and Decision Making, 3(10).
6. Alia Arshad, Kanwal Ameen , (2015). Usage patterns of Punjab University Library website: a transactional log analysis study, The Electronic Library, 33(1), 65 – 74.
7. Albornoz, S. et al. (2007), "Information searching and library use by researchers in the humanities at the University of Argentina", Revista Interamericana de Bibliotecologia, 30(1),73-92.
8. Amjad, A.., Ahmed, S.., Naeem, S.B., (2013). Use of Electronic Information Resources among Research Scholars in the Islamia University of Bahawalpur, Pakistan New Review of Academic Librarianship, 19(3), 316-328.

9. Anaraki, L.N. , Babalhavaeji, F., (2013). Investigating the awareness and ability of medical students in using electronic resources of the integrated digital library portal of Iran: A comparative study. *Electronic Library*, 31(1), 70-83.

10. Appleton, L. (2004). The use of electronic books in midwifery education: the student perspective. Health information and libraries journal, 21(4), 245-252.

11. Asunka, Stephen. (2013). The viability of e-textbooks in developing countries: Ghanaian University students perceptions, Open Learning, 28(1), 36-50.

12. Asemi, A., and Riyahiniya, N., (2007). Awareness and use of digital resources in the libraries of Isfahan University of Medical Sciences – Iran, Electronic Library, 25(3),316-327.

13. Atakan, Cemal. Atilgan, Dogan., Bayram, Ozlem., and Arslantekin, Sacit., (2008). An Evaluation of the Second Survey on Electronic Databases Usage at Ankara University Digital Library, The Electronic Library, 26(2), 249-259.

14. Barbara Bren, Beth Hillemann, Victoria Topp (2015), Effectiveness of hands-on instruction of electronic resources, Research Strategies, 16(1), 41-51.

15. Boumarafi, B., (2010). Electronic resources at the university of Sharjah medical library: An investigation of students' information-seeking behavior, Medical Reference Services Quarterly, 29 (4), 349-362.

16. Celeste Feather (2016). Electronic Resources Communications Management: A Strategy for Success, the Electronic Resources and Libraries Conference, Georgia Institute of Technology

17. Chopkar, H.D. and Khaparde, V.S. (2012), "Information Seeking Behaviour of Research Scholars of Biological Science Department of Dr. Babasaheb Ambedkar Marathwada University: A Study", International Journal of Information Dissemination and Technology 1 (4), 241-243.

18. Coates, H. (2013).Academic historians in Canada report both positive and negative attitudes towards E-books for teaching and research. Evidence Based Library and Information Practice. 8(4), 148-150.

19. Debbie Malone, Carol Videon (2015). Assessing undergraduate use of electronic resources: A quantitative analysis of works cited, Research Strategies, 15(3), 151-158.

20. Dhanavandan, S., (2014). Knowledge and utilization of e-resources and services in Annamalai University Library, International Journal of Information Dissemination and Technology.

21. Dhanavandan,S., Mohamed Esmail,S. andSivaraj,S.,(2000). Use of digital resources among Engineering Professionals in Cuddalore District, Tamil Nadu, Indian Journal of Information Science and Services, 3(1), 36-1.

22. Drachen, T.M. and Asger, V.L. (2011), "Information behaviour and practices of PhD students Appendices". Retrieved on 16.01.2016 from https://halshs.archives- ouvertes.fr/hprints- 0599034/file/Information_behaviour_and_practices_of_PhD_stud ents.pdf

23. Doney, L., Barlow, H., West, J.,(2005). Use of libraries and electronic information resources by primary care staff: outcomes from a survey. Health infor*mation and libraries journal*, 22(3), 182-188.

24. Ellis, D., Cox, D., and Hall, K., (1993). A Comparison of the Information-Seeking Patterns of Researchers in the Physical and Social Sciences. Journal of Documentation, 49(4), 356-369.

25. Fasola, O.S. and Olabode, S.O. (2014), "Information seeking behaviour of students of ajayi crowther university, oyo, Nigeria", Brazialian journal Information Science, Marília 7(2).47-60.

26. Francis, H., (2005). The information seeking behaviour of Social Science Faculty at the University of the West Indies, St. Augustine Campus, Journal of Academic Librarianship, 31(5), 67-72.

27. Frost, et al., "Some Economic and Legal aspects of Open Source Software, May 7th 2005

28. Gopinath, S. A. (2017). Perception and Use Electronic Information Resources Among the Academic Community: Acase study. International Journal of Information Dissemination and Technology, 7(2), 142-145.

29. Gowda, V and Shivalingaiah, D. (2009), "Attitude of research scholars towards usage of electronic information resources: A survey of university libraries in Karnataka", *Annals of Library and Information studies*, 56(3), 184-191

30. Gowda, Vasappa., and Shivalingaiah, D., (2009). Attitude of Research Scholars towards Usage of Electronic Information Resources: A Survey of University Libraries in Karnataka, Annals of Library and Information Studies, 56(3), 184-191.

31. Govindaraju, Nemani., (2010). Use and user awareness of E-resources in Andhra University library: A study, PEARL: A Journal of Library and Information Science, 4(3), 183-188.

32. Halima SadiaEgberongbe (2011).The use and Impact of Electronic Resources at the University of Lagos, Library Philosophy and Practice,

33. Haneefa, K. (2007). Use of ICT based resources and services in special libraries in Kerala. Annals of Library and Information Studies, 54(1), 23-31.

34. Holleman, Curt (2016). Electronic Resources: Are Basic Criteria for the Selection of Materials Changing?, Library Trends Spring,694.

35. Jonathan P. Egle, David M. Smeenge, Kamal M. Kassem, Vijay K. Mittal (2015). The Internet School of Medicine: Use of Electronic Resources by Medical Trainees and the Reliability of those Resources, Journal of Surgical Education, 72(2), 316– 320.

36. Joteen Singh, R.K., Devi, Th. Madhuri., and Raychaudhury, Arup., (2009). Use of Internet Based E-Resources at Manipur University: A Survey, Annals of Library and Information Studies, 56(1), 52-57.

37. Joanna Duy, Liwen Vaughan (2015), Usage data for electronic resources: A comparison between locally collected and vendor-provided statistics, The Journal of Academic Librarianship, 29 (1),16-22.

38. Kebede, G. (2002). The changing information needs of users in electronic environments. The Electronic Library, 20(1), 19 - 21.

39. Khan, A.M. (2012), ".Use of Information Sources by Faculty Members, Research scholars and Students of the Faculty of Commerce, AMU, Aligarh: A Survey", Library Philosophy and Practice (e-journal). retrieved from: http://digitalcommons.unl.edu/cgi/viewcontent.cgi?article=1895&context=libphilprac.

40. Kindilchie, A.I., and Samarraie, I.F., (2008). Interaction and impact of electronic information resources on Qatar university faculty, Libri, 58(4), 281-293.

41. Kumar, G.T., and Kumar, B.T. Sampath., (2008). Use of electronic information sources by the academic community: A comparative study. 6th International CALIBER-2008, Allahabad: University of Allahabad, 684-692.

42. Lakan, E.L., (2008). Availability and Utilization of Electronic Information Databases by Staff of the Agricultural Complex, Ahmadu Bello University, Zaria, Samaru Journal of Information Studies. 8(1), 44-53.

43. Lohar M.S and Roopashree T.N (2015). Use of Electronic Resources by Faculty Members in B.I.E.T, Davanagere: A Survey, Science, 43 (1), 201.

44. Madhusudhan, M., (2010).Use of electronic resources by research scholars of Kurukshetra University, Electronic, 28(4), 492-506.

45. McClanahan, K., Wu, L., Tenopir, C., and King, D.W., (2010). Embracing change: Perceptions of e-journals by faculty members, Learned Publishing, 23(3), 209-223.

46. Mahapatra, R. K. (2017). Use of e-Resources among Social Scientists in Selected Institutes in Bhubaneswar: Astudy. International Journal of Information Dissemination and Technology, 7(2), 102-106.

47. Mostofa, S.M. (2013), A Study of Information Needs and Seeking Behaviour of Faculty Members of Darul Ihsan University in Bangladesh", Library Philosophy and Practice (e-journal). 983. http://digitalcommons.unl.edu/libphilprac/983.

48. Nwagwu, W.E.ab.,Segilola, M.O.a., (2013). Information needs information sources and information seeking behaviours of engineers in breweries in Nigeria. Turkish on*line journal of education*, 63(2), 159-178.

49. Ngozi, O V, Uche, EU and Ejiro, AK (2015), Information seeking behaviour of faculty members in a Nigerian University, International Journal of Academic Research and Reflection, 3(2), 95-102.

50. Nzivo, Charles N., and Chafe, Chen., (2013). International Students' Perception of Library Services and information resources in Chinese academic libraries, Journal of Academic Librarianship, 39(2), 129-137.

51. Okello-Obura.C., (2010). Assessment of the problems LIS Postgraduate students face in accessing e-resources in Makerere University, Uganda, Collection Building, 29 (3), 98-105.

52. Ollé, Candela., and Borrego, Ángel., (2010). Librarians' perceptions on the use of electronic resources at Catalan academic libraries: results of a focus group, New Library World, 111(1/2), 46-54.

53. Pandey,R., and Mehta,S., (2002). Awareness if Educational Technologies in open learning system by Target group. Indian Journal of Social Research, 43(3), 183-189.

54. Rafiq, M. and Ameen, K. (2009), "Information seeking behaviour and user satisfaction of university instructors: A case study", Library Philosophy and Practice, Retrieved on 26.12.2015 from www.webpages.uidaho.edu /~mbolin/rafiq-ameen.htm.

55. Pradeep Mittal and BonuBala (2013).Use of e-Resources in Universities, International Journal of Innovative Research in Computer and Communication Engineering, 1(6)

56. Sami, Lalitha K; and Iffat, Rabia. (2009). Use of Electronic Information Services in Research Libraries. SRELS Journal of Information Management, 46(1), 43-50.

57. Santosh A Navalur, Balasubramani.R, Ashok Kumar.P, (2012). Usage of E-resources by Faculty, Research Scholars and PG Students of Bharathidasan University: A Study, Journal of Advances in Library and Information Science, 1(4), 165-172.

58. Sasireka, G., Balamurugan, S., Gnanasekaran, D., (2011). Use of E-resources in digital environment among engineering institutions in Tamil Nadu (India): An empirical study. European Journal of Scientific Research, 60(3), 326-333.

59. SeemaVasishta., (2014). Use pattern of E-resources by research scholars and faculty: A survey of technical university libraries in North India, International Journal of Library and Information Studies, 4 (1).

60. Selvi,M., and Dhanavandan,S., (2012). A Study on the Utilization of Interne Resources and Tools by Basic Medical Science Professionals. Journal of Emerging Trends in Computing and Information Sciences, 3(12), 1634-1637.

61. Sheeja, N., (2010). Science vs Social Science: A study of information-seeking behavior anduser perceptions of academic researchers, Library Review, .59(7), 522- 530.

62. Shuling, W., (2007).Investigation and analysis of current use of electronic resources in university libraries, Library Management, 28(1-2), 72-88.

63. Sudhier.K.G., and Seethalekshmi, K P., (2011). Use of E-resources by the students and Researchers of Faculty of Arts, University of Kerala, International Journal of Information Dissemination and Technology, 1(3),

64. Sujatha, H.R., and Mudhol, Mahesh V., (2008). Use of Electronic Information Sources at the College of Fisheries, Mangalore, India. Annals of Library and Information S*tudies*, 55(3), 234-245.

65. Swain, D.K., Panda, K.C., (2009). Use of electronic resources in business school libraries of an Indian state: A study of librarians' opinion. Electronic Library, 27(1), 74-85.

66. Thanuskodi, S. (2010). Use of internet and electronic resources for agricultural science information: A case study. Social Sciences, 5(4), 364-367.

67. Thelwall, Mike et al. (2001). Custom interfaces for advanced queries in search engines. ASLIB Proceedings, 53(10), 413-422.

68. Thangaraja,K., Karthikeyan,C.,Asokan, M., and Rajasekaran, R., (2008). Study on the Utilization behaviour of online journals by the students of TNAU, The Madras Agricultural Journal, 95(1-6), 114-119.

69. Thanuskodi, S., (2011). Usage of electronic resources at Dr T.P.M. Library, Madurai Kamaraj University: A case study, DESIDOC Journal of Library & Information Technology, 31(6), 437-445.

70. Trivedi, Mayank., and Joshi, Anuradha., (2009). Usage of Electronic Journals (E- Journals) Versus Print Journals by Healthcare Professionals in H M Patel Centre for Medical Care and Education (HMPCME). Journal of Health Informatics in Developing Countries, 3(1), 13-16.

71. Vakkari, P.,(2008).Perceived influence of the use of electronic information resources on scholarly work and publication productivity, Journal of the American Society for Information Science and Technology, 59(4), 602-612.

72. UmmeHabiba and Salma Chowdhury (2012), Use of electronic resources and its impact: A Study on Dhaka University Library users, The eastern *Librarian,* 23(1), 74- 90.

73. Weiler, Angela. (2005). Information-Seeking Behavior in Generation Y Students: Motivation, Critical Thinking, and Learning Theory, Journal of Academic Librarianship, 31(1), 46-53.

74. Yan, Yalan., Zha. Xianjin.,and Xiao, Zhiliang., (2013). Exploring Users' Perceptions of Conventional and Unconventional electronic resources, Serials Review, 39(2), 105- 113.

75. Zha, Xianjin., Li, Jing., and Yan, Yalan.,(2012). Understanding Usage Transfer from Print Resources to Electronic Resources: A Survey of Users of Chinese University libraries, *Serials Review*, 38 (2), 93-98.

3. Thesis Abstract

1. Natarajan, K., Suresh, B., Sivaraman.P, and Sevukan, R., (2010). Use and user perception of electronic resources in Annamalai University: A case study, *Annals of Library & Information Studies*, 57(1), 59-64.

4. Websites

1. Bruce Perens, "The Debian Free Software Guidelines" June 1997. www.opensource.org/docs/definition.php.

2. https://en.wikipedia.org/wiki/List_of_institutions_of_higher_education_in_Tamil_Na du

3. http://www.msuniv.ac.in/MSULibrary/index.html

4. http://www.periyaruniversity.ac.in/Library_Home/index.html

5. http://library.bdu.ac.in/

6. http://www.b-u.ac.in/Home/Library

7. http://www.motherteresawomenuniv.ac.in/

8. http://www.mkulibrary.org/

9. ICDL 2004. www.teriin.org/events/icdl/background.htm.

10. Khan, S.A. and Bhatti, R and Khan, A.A. (2011), "User of ICT by Students: A Survey of Faculty of Education at IUB", Library Philosophy and Practice (e-journal). Retrieved on 16.01.2016 from http://digitalcommons.unl.edu/libphilprac /677/

Interview Schedule

Please make a tick mark against the relevant answer

Section - I Demographic Characteristics of Library Users

1. **Name**	:		
2. **Gender**	:	1) Male	2) Female
3. **Age**	:		
		1) Upto 30	2) 31-40
		3) 41-50	4) 41 and above
4. **State**	:	1) Tamil Nadu	2) Other state
5. **Residential Sector**	:	1) Rural	2) Urban
6. **Marital Status**	:	1) Married	2) Unmarried
7. **Educational Qualification**	:	1) Postgraduate	2) M.Phil
		3) Ph.D	4) Post Doctorate
8. **Status of users**	:		

1) Research Scholars: i) M.Phil ii) Ph.D

2) Faculty Members: i) Assistant Professor ii) Associate Professor iii) Professor

9. Religion : 1) Hindu 2) Muslim 3) Christian 4) Others

10. Name of the University : ....

11. School/Faculty : ....

12. Department : ....

Section - II Use of University Library

13. How often do you visit to the University library?

1) Daily 2) Once in a Week 3) Twice in a week
4) Monthly 5) Rarely

14. How much time do you spend in the library for single visit?

1) One hour 2) 1-2 hours 3) 2-3 hours
4) 3-4 hours 5) More than 4 hours

15. What is your purpose of visiting the library?

1. Read Newspapers and Read Periodicals
2. Prepare for Competitive Exams
3. General Knowledge and Employment News
4. Entertainment
5. Borrow Books
6. Academic Improvement and Research purpose
7. E-Resources and Others

16. Do you think that you have developed your general knowledge after going to the Library?

1) Strongly Disagree 2) Disagree 3) No comments
4) Agree 5) Strongly Agree

Section - III Electronic Information Sources

17. Where do you mainly access the electronic information?

1) Department Library 2) Central Library 3) Hostel
4) Home 5) Internet Cafe

18. What is your level of using the Electronic information resources?

1) Beginner 2) Average 3) Expert

19. Frequency of using the Electronic information resources.

1) More Frequently 2) Frequently 3) Rarely

20. How many hours do you spend for electronic information per day.

1) Less than 30 mts 2) 30 mts-1 hour 3) 1-2hours
4) 2-3 hours 5) More than 3 hours

21. How did you learn the use of E- resources?

1) Guidance from friends 2) Trial & Error Method/Self Instruction
3) Formal training 4) By Attending Courses
5) From teachers 6) From Librarians

22. Please tick the type of E-resources accessed.

1) E-Books 2) E- Journals
3) E- News Papers 4) E-Thesis and Dissertation
5) E-Magazines 6) CD-ROM databases
7) Online Databases 8) Others any --------------------------

23. What is the purpose of using the Electronic information resources?

1) To keep abreast with the latest development
2) For study/ research
3) For career development

4) For teaching preparation
5) For writing books/ research articles
6) For communication (e-mail)
7) Professional Development
8) E-shopping
9) Entertainment
10) Others

24. Rating the features of Electronic information.

S.No	Features	Poor (1)	No comments (2)	Satisfaction (3)	Good (4)	Very good (5)
1	Availability					
2	Accessibility					
3	Accuracy					
4	Reliability					
5	Ease of use					
6	Flexibility					
7	Stability					
8	Timeliness					
9	Uniqueness					

25. Do you accept that accessing information in a Digital Library

1) Saves more physical space.
2) Library services shift from 'just in case' to 'just in time'
3) Access from anywhere at any time.
4) Availability of information to all users simultaneously
5) Easy retrieval
6) No Maintenance
7) Enables networking of various internal resources
8) Networking of other libraries
9) Others

26. To what extent are you satisfied with the following electronic information resources

S.No.	Sources	Dissatisfied (1)	No Comments (2)	Somewhat Satisfied (3)	Satisfied (4)	Highly Satisfied (5)
1)	Internet Sources					
2)	Subscribed Databases					
3)	Open source Databases					
4)	E- Books					
5)	E- Journals					
6)	CD-ROM Databases					
7)	A/V Materials					
8)	E-Catalogue					
9)	Ask a Librarians					
10)	Others					

27. Do you have access to UGC- eSS E-Journals?

1) Yes 2) No

28. For What purpose are you using UGC- eSS E- Journals?

1) Research Work
2) Publishing articles/ books
3) Keeping up-to-date information
4) Finding relevant information in the area of specialization
5) UGC NET

29. Please indicate the problems faced by you in using UGC- eSS E- Journals.

1) Lack of knowledge to use
2) Lack of sufficient nodes in university

3) Content searching is difficult as compared to search Engine
4) Slow Internet Bandwidth
5) Technical Problems (Server Down)
6) Frequent Power cut
7) Lack of assistance by library staff
8) Restricted access
9) Others

30. UGC-ESS journals have developed your academic skills.

1) Strongly Disagree 2) Disagree 3) No comments
4) Agree 5) Strongly Agree

31. What is your level of satisfaction in using UGC-eSS consortium?

1) Very Dissatisfied 2) Dissatisfied 3) No comments
4) Satisfied 5) Very satisfied

32. Difficulties encountered in accessing E-journals

S.No	REASONS	Never (1)	Rarely (2)	Sometimes (3)	Most often (4)	Often (5)
1	Required Password not Known					
2	Access is restricted to Campus Only					
3	The E-Journal Interface is difficult to Navigate					
4	The Library does not have subscription to Journals required					

33. Mention the level of Satisfaction with Electronic sources for research purpose.

1) Not helpful 2) No comments
3) Helpful 4) Most helpful

34. To What extent do you use the electronic resources and printed materials in your own work?

1) Use Only Electronic Resources
2) Use both Electronic and Printed Materials But Mostly Electronic
3) Use Electronic and Printed Materials Equally
4) Download Electronic But Convert to Print For Reading
5) Use mostly Printed Materials
6) Others

35. Perceived advantages of E-resources.

1) Multiple Choices of Formats (PDF, PS, RTF, DOC, HTML and others)
2) Provides Faster and Reliable Communication
3) Multi User Access
4) Users from their Desktop can Access Electronic Information Resources
5) Access to a wide range of information.
6) Easier access to information within a few minutes.
7) Others

36. Level of satisfaction of information obtained from library using E-resources.

1) Dissatisfied 2) Partially satisfied 3) No Comments
4) Satisfied 5) Highly satisfied

37. Respondents' satisfaction on E-resources.

Kindly rate the satisfactory level of getting the following E-resources: (1 is the lowest, 9 is the highest score)

S.No	E-resources	Score
1	E-Books	
2	E- Journals/E-magazine	
3	E-Dictionaries	
4	E-Encyclopedia	
5	E-Theses	

S.No	E-resources	Score
6	E- Newspapers	
7	Online Databases	
8	E-Mail, Discussion Groups	
9	E- Research Reports	

38. User satisfaction about the E-resources available.

S.No	Options	Poor (1)	No comments (2)	Satisfaction (3)	Good (4)	Very good (5)
1	Quality					
2	Currency					
3	Authenticity					
4	Easy to retrieve					
5	Downloading speed					
6	Up to dateness					

39. Indicate the importance of using E-resources

S.No	Features	Not Important at all (1)	Not Important (2)	Somewhat Important (3)	Important (4)	Very Important (5)
1	Research/Project					
2	Lesson Plans					
3	Accessing online databases					
4	Publishing					
5	Downloading programs					
6	Placements					
7	Chatting					
8	Entertainment					
9	Others					

40. The Library provides adequate training and satisfaction of the electronic resources

S.No	Features	Strongly Disagree	Disagree	No comments	Agree	Strongly Agree
1	Library provides adequate training on how to use the electronic resources					
2	Electronic information develops academic skills					
3	Library provides adequate electronic resources					
4	level of satisfaction while using electronic information resources					

<u>Section - IV Electronic Information Services</u>

41. What are your purposes of using electronic Information Services?

S.No	Features	Strongly Disagree	Disagree	No comments	Agree	Strongly Agree
1	To know the availability of materials					
2	To get study/research related materials					
3	To get material for class assignments					
4	To get literature for research					
	To get the print out of required materials					

42. What type of problems are you facing in accessing electronic information services?

S. No	Problems	Strongly Disagree	Disagree	No comments	Agree	Strongly Agree
1	Time Consuming					
2	Slow Speed					
3	Not easy to search					
4	Not familiar with the systems					
5	Not enough terminals to access information					
6	Lack of trained staff					

43. How do you rate the value/ importance of these electronic reference services?

S.No	DigitalReference Format	Not Important at all (1)	Not Important (2)	Somewhat Important (3)	Important (4)	Very Important (5)
1	E-mail Reference					
2	Web forms					
3	Ask-A Librarian Service					
4	Online Chat Reference					
5	Social Media					

44. The following is the list of electronic Information Services available in the library.

How frequently do you use these services?

S. No	Electronic Reference Format	Frequency				
		Never (1)	Rarely (2)	Sometimes (3)	Frequently (4)	Most Frequently (5)
1	OPAC/ Web OPAC					
2	CD ROM Database Service					
3	Current Content Service					
4	New additions to library (E- Alerts)					
5	E-Circulation					
6	Online Reference Service					
7	Current Awareness Service(CAS)					
8	Selective Dissemi- nation Services (SDI)					
9	Video & Teleconferencing Service					
10	Online Reservation of Reading Materials					
11	Newspaper Clipping Service					
12	Electronic Feedback					

Made in the USA
Monee, IL
05 November 2026

67807954R00083